Samira Soualem

Salinity tolerance mechanisms

Samira Soualem

Salinity tolerance mechanisms

ScienciaScripts

Imprint
Any brand names and product names mentioned in this book are subject to trademark, brand or patent protection and are trademarks or registered trademarks of their respective holders. The use of brand names, product names, common names, trade names, product descriptions etc. even without a particular marking in this work is in no way to be construed to mean that such names may be regarded as unrestricted in respect of trademark and brand protection legislation and could thus be used by anyone.

Cover image: www.ingimage.com

This book is a translation from the original published under ISBN 978-613-9-52741-0.

Publisher:
Sciencia Scripts
is a trademark of
Dodo Books Indian Ocean Ltd. and OmniScriptum S.R.L publishing group

120 High Road, East Finchley, London, N2 9ED, United Kingdom
Str. Armeneasca 28/1, office 1, Chisinau MD-2012, Republic of Moldova, Europe
Printed at: see last page
ISBN: 978-620-8-26716-2

SALINITY TOLERANCE MECHANISMS

SUMMARY

Alriplex halimus L. is an important forage plant in arid Mediterranean areas. The aim of this study is to identify, within the natural genetic diversity of a population, the types that can behave optimally under certain saline conditions. We focused our observations on the in vitro behaviour of plants watered with a saline solution composed of NaCl +$CaCl_2$ at 400 and 600 meq and with seawater diluted to 50% and undiluted. Various morphological, physiological, biochemical and anatomical parameters are involved in adaptation to salinity. The results show that the indicators of the plant's water status are sensitive to variations in the salt content of the substrate. In addition, the use of two types of substrate with increasing salt concentrations caused structural and biochemical variations in the plants tested. However, the behaviour of the plants in relation to the origin of the salt was not always similar, and this reaction needs to be taken into account if we are to gain a better understanding of the complexity of the mechanisms that enable plants to cope with stress.

TABLE OF CONTENTS

INTRODUCTION

Salinity and drought are major constraints which considerably limit plant productivity over 40% of the land surface, particularly in Mediterranean regions (M'barek et al., 2001).These two natural constraints have altered the stability of ecosystems (Lieth et al., 1997) and are largely responsible for soil desertification. Some spontaneous species have disappeared and others are threatened with extinction.More than 15% of Algeria's surface area is occupied by saline soils. They extend discontinuously from north to south, parallel to the coast.Soil salinisation in these regions is linked not only to climatic conditions but also to the often poorly controlled use of irrigation. High levels of sunshine and low rainfall cause dissolved salts to accumulate on the surface (Levigneron et al., 1995). The accumulated salts can limit or completely halt plant growth as a result of an increase in the osmotic pressure of the environment or the specific toxic effect of certain elements (Chretien, 1992).This phenomenon has also been accentuated by the excessive use of fertilisers, leading to a process of secondary salinisation (Hamdy, 1999). In crops grown under cover, very localised fertilisation and irrigation lead to an exaggerated increase in the concentration of salts in the growing media.The introduction of salinity-tolerant fodder shrubs is one of the techniques that can be used to make the most of these marginal soils, and Atriplex has good tolerance of the unfavourable environmental conditions (Kinet et al., 1998).For example, in a liquid synthetic medium, Atriplexhalimus L tolerates concentrations of sodium chloride (NaCl) close to those of seawater (600mM). This species is an important fodder reserve in North Africa and the Middle East, consumed by sheep and camelids, especially in times of famine. It is rich in protein (10 to 20% of dry matter) and has a highly developed root system that fixes the upper layers of the soil and can be used to combat desertification (Ben Ahmed et al. , 1996).Atriplex exhibits considerable morphological polymorphism in the size and shape of its leaves, fruiting valves and seeds, as well as polymorphism in biomass production. It would therefore be interesting to improve steppe production based on Atriplex halimus L. (Ben Ahmed et al., 1996), by developing the main strategies for repopulating degraded areas using resistant homogeneous populations adapted to each region (Kinet et al., 1998).There are many indications in plants of considerable genetic potential for tolerance to environmental stresses (Mallek-Maalej et al., 1998). They have different degrees of resistance to salinity and water stress, well established on the basis of agronomic criteria (e.g. yield). Knowledge of this interspecific variability is based essentially on farming practices, and to a lesser extent on systematic comparative field trials (Maas, 1986).More interesting is the intraspecific variability that plants display in terms of tolerance to abiotic stresses. This makes it possible to select resistant genotypes. This creationvarietal development requires tests that are sufficiently simple and rapid to enable large-scale sorting prior to yield trials in the field. Stress resistance is highly dependent on the plant's stage of development, cultivation techniques and climatic and soil conditions (Digby et al., 1999). It cannot therefore be defined or predicted in absolute terms, but must be considered as a relative parameter. Nevertheless, several physiological and biochemical criteria can be retained, at the level of the cell or the whole plant, and used to discriminate between species or varieties for stress resistance (Tal et al., 1984).The fact that resistance to salt or water stress can be acquired by cells or tissues in culture implies that this physiological behaviour is dependent on cellular parameters (Khrais et al., 1998).However, a number of studies show that the ability to withstand the salinity of the environment is associated with a precise pattern of transport integration at the level of the entire plant. This suggests that criteria based on the measurement of cellular parameters can be used to detect elementary limiting factors, which alone confer a high degree of sensitivity to stress, leading to the screening of the best genotypes for optimising the performance of stressed plants (Beatriz et al., 2001).The aim of this study is to quantify the effects of different salt concentrations in seawater and a salt solution composed of $NaCl^{+}$ $CaCl_2$ on a number of physiological, morphological, anatomical and biochemical parameters.In the first part we will present the data acquired on the physiological parameters, mainly variations in the plant's water status, osmotic potential and the parameters linked to plant transpiration under salt stress.The second part will focus on the biochemical parameters to be monitored.through leaf analysis of sugars and wax levels.The third part focuses on morphological characteristics, in particular the root system and aerial part.Finally, anatomical observations of the leaves and roots of stressed plants complete this work.

CHAPTER I
BIBLIOGRAPHICAL DATA

I - ABIOTIC STRESSES

The physico-chemical stresses to which plants are subjected can be grouped into a few broad types: drought (water stress) and nutrient deficiencies, chemical poisoning, excessive or insufficient light, and excessively low or very high temperatures.

1- Drought

Drought can be defined as a negative deviation sufficient to cause a significant drop in agricultural production (Zair, 1991).According to Deressac in 1992, there is drought as soon as water becomes limiting factor for growth and yield.The problem of drought is, and will continue to be, a major issue in agro-economic chronicles. Water deficiency, which when persistent can lead to severe dehydration of plant tissues, is one of the most common stresses affecting crops. Research into the mechanisms by which plants adapt to water deficit is of paramount importance. The mechanisms of resistance to drought are highly complex; they may involve morphological, physiological and biochemical factors at different levels of plant organisation (Reda Tazi et al., 2003).According to Levitt in 1980, two types of drought adaptation mechanism have been envisaged: the avoidance mechanism and the tolerance mechanism. Neither of these characteristics alone can be used in selection for resistance to a stress; rather, it is their combination that seems to confer on the plant its ability to tolerate such a condition. According to Turner in 1986, the processes which are linked to the growth and development of the plants are influenced by lack of water and salinity.

2- Salinity

The problem of salinity in agriculture is becoming increasingly important in various parts of the world, especially in the arid zones of the Third World, as well as in certain countries developing modern agriculture where soil quality has often deteriorated as a result of excessive irrigation (Christophe and Michel, 2000).

a- Definition of salinity

Salinity is defined as the pedological process whereby the soil becomes abnormally enriched in soluble salts, thus acquiring a saline character (Eilers et al., 1995). Thus, saline soils are those characterised by the presence of an excess of exchangeable sodium ions in the profile (Schut, 1996).

According to Aubert in 1975, sodic soils, also known as saline soils or halomorphic soils, are characterised by a high content of soluble salts, distributed over all or part of the profile. In particular, the influence of the sodium ion results in the degradation of the structure of one of the horizons, leading to the formation of a crust on the surface of the soil; this crust is powdery in nature and the soil is then colonised by halophytic vegetation (Schut, 1996).

According to Loze and Mathieu in 1990, a soil, irrigation water or nutrient solution is salty when it contains abnormally high concentrations of chlorides, sulphates, carbonates or bicarbonates of sodium, calcium or magnesium.The formation of halomorphic soils is conditioned by the presence of the Na ion^{+} , which, when sufficiently abundant in the soil, gives it special properties. It can exist in two forms, with different properties: the neutral saline form, generally chlorides of which has not any properties and the exchangeable form, linked to the Omplex absorbent which on the alkalinizes the solution (Duchaufour, 1983).

b-Classification of salty soils :

Halomorphic soils can be divided into two main types:

• Saline soils with a calcium complex:(calcium solantchacks) :

These soils are found in steppe and sub-desert regions (La coste et al., 1980), and contain sodium salts, Ca salts and Mg salts (Duchaufeur, 1988).), containing not only sodium salts but also Ca^{+2} and Mg^{+2} (Duchaufour, 1988). The salts often accumulate on the surface during periods of high evaporation in the dry season, causing white efflorescence to form on the surface (Duchaufour, 1983).These soils are subject to very high osmotic pressure, preventing the absorption of water and other ions, resulting in real physiological drought (Duchaufour, 1968). They offer little exchangeable sodium, generally less than 15%; alkalinisation results in PH > 8.5 and an EC < 4 mmhos/cm at 25°C (Duchaufour, 1988).

• Saline soils with a sodium complex (sodium solantchack) :

This type is found on the seashore, where the water, which is very rich in sodium ions, contains little alkaline earth Ca^{+2} , Mg^{+2} , in which case the sodium ion represents 15% of the EC value (Duchaufour, 1988). However, due to the high concentration of sodium in the water at all times of the year, the influence of partial saturation of the complex with sodium is only attenuated. The PH does not exceed around 8.5. According to the same author, sodium salts (NaCl, Na_2SO_4) rising from the water table accumulate on the surface as in the previous case; the structure tends to degrade and becomes powdery.

c- Origin of salinity

• primary origin

This salinity results from the alteration of the saline parent rock b y erosion factors. The dissolution, by run-off water, of sedimentary rocks which are rich in chlorides, sulphates and carbonates, thus contributing to the salinisation of soils (Duchaufour et al., 1979). Alteration of the parent rock, which supplies the salts responsible for primary salinisation, is caused by rainwater, which is often acidic. (H_2CO_3) but also by physical agents (Aubert and Boulaine, 1980).

• secondary origin

This process does not involve the bedrock, but can be caused by irrigation water, which is loaded with salts, or by excessive chemical fertilisation (Mouhouch e and Boulassel, 1999). In addition, in irrigated soil, capillary rise occurs during intervals between irrigations and during a fallow period when there is no downward movement due to water percolation (Van-Hoorn, 1995).According to Eilers et al in 1995, the recharge zone receives rainfall, the water penetrates the soil and if it is not absorbed by the plants, it reaches the groundwater, which dissolves the salts and transports them to the discharge zones where the groundwater supply raises the water table to the capillary fringe like a sponge, causing the water to rise to the surface of the soil, which evaporates and the salt accumulates.

d-The main soluble salts and their characteristics in soil

The soluble salts that accumulate in salty soils are essentially made up of Ca^{++} , Mg^{++} ,Cl^- ,Na^+ and SO_4^{--} ions.K ions$^+$, HCO_3^- ,CO^-_3 , are less abundant,the proportion of these ions varies considerably in different soils (Dommerguez and Mangenot,1970).

Carbonates

They result from the dissolution of carbonic acid (H_2CO_3) in water, which may be of atmospheric or biological origin. Carbonate ions are found in soils with a PH of 9 or more, mainly in the form of :

- magnesium carbonates ($MgCO_3$) with very high solubility.
- sodium carbonates (Na_2CO_3) give solutions with a very high PH 9 -12, they are highly toxic both because of their alkalinity (as is the case for bicarbonates($NaHCO_3$)and carbonates of potassium carbonates (K_2CO_3) (Sylvia,1982 in Gasmi,1989).

Sulphates:

Soluble sulphate-based salts are :

- Magnesium sulphates ($MgSO_4$) are found in groundwater and are highly soluble.
- sodium sulphates (Na_2SO_4): their solubility depends on temperature and their action depends on the season (Duchaufour et al., 1979).

Chlorides

Soluble salts based on

- calcium chlorides ($CaCl_2$): rare in soils, they are highly soluble and react with Na_2SO_4 or Na_2CO_3 to give $CaSO_4$ or $CaCO_3$ which are not very soluble.
- magnesium chloride ($MgCl_2$): more common than $CaCl_2$ especially in salts with a high salinity, they are very soluble.
- Sodium chlorides (Na Cl) are widespread, highly soluble salts typical of salty soils (as are potassium chlorides (K Cl) (Eilers et al., 1995).

e- Distribution of salinity

Large areas of the world's land are naturally saline, or undergo secondary salinisation as a result of irrigation or in hot climates with evaporating atmospheres. This poses a serious threat to agricultural

development in many countries (Houchi and Coudret, 1994).
The areas lost to agriculture each year as a result of secondary salinisation vary between 10 and 20 million hectares per year, depending on the author (Cheverry, 1995), but it is above all the very large increase in irrigated areas, often accompanied by poor water management, that is responsible for the expansion and diversification of saline environments. Of the 280 million hectares now irrigated worldwide :
-75.6 million hectares, or 27%, are already affected by salinity.

-140 million hectares, i.e. 50%, are threatened in the long term by climate change.salinity (Larsam, 1995).

An estimated 1 billion hectares of land worldwide are affected by salinity, mainly in arid and semi-arid regions.

Resistance to salinity

The expression of plants to salinity results in both hydric stress due to the osmotic effects of salt, and chemical stress mainly due to the toxic effects of sodium. Assessing the response to salinity of a given species or variety is therefore not an easy task, due to the influence exerted by a number of factors that are often difficult to control (Shannon et al., 1985; Amrar, 1993).

Depending on the degree of salinity of the surrounding environment, plant species can be grouped into halophytes and glycophytes.
Glycophytes are plants that can live in a salty environment, but grow better in a salt-free environment.Halophytes, on the other hand, are plants that need salt to grow. They are therefore plants that always live in a salty environment (Hubac, 1990).Higher plants exposed to salinity have developed physiological mechanisms that enable them to live in halomorphic soil.

a- Avoidance

- **exclusion**

This phenomenon is well known in halophytes, which on sunny days secrete salts in the form of visible hoppers on the surface of their leaves (Batamouny, 1993).According to Schubert et al 1980 quoted by Amrar in 1993, sodium exclusion occurs at two levels. Firstly, through a mechanism that restricts the uptake of Na^+ by plant roots, in

particular by reducing the permeability of root cells to Na ions^{+} and secondly through the exclusion of Na^{+} by the plasma membrane of conducting tissues during the K^{+} Na^{+} exchange.
The exclusion of salts enables an ionic balance to be maintained in the presence of a high concentration of monovalent cations, following selective uptake of Ca^{++} on the plasma membrane (Alem et al., 2005).
Under conditions of salt stress, there is a decrease in the calcium/sodium ratio in the root environment, which affects the properties of the membrane, leading to a breakdown in membrane integrity (Kramer et al., 1995; Kinraide, 1998).

This favours an increase in the Na^{+} content inside the cells, which could change enzymatic activity, resulting in changes in metabolism, K uptake^{+} , cell division and may even affect stomatal regulation (Epstein, 1998). Furthermore, the addition of Ca^{+2} to the root environment in the presence of salt maintains or increases the selective uptake of potassium, which will prevent the deleterious effects of excess Na^{+} (Epstein, 1998 ; Da Silva et al., 2003).
Another role attributed to the supplementary addition of Ca^{2+} is its contribution to osmotic adjustment by improving the accumulation of dissolved organic bodies (Colmer et al., 1996; Girija et al., 2002).
It has been shown that a Ca-binding protein^{++} , homologous to the regulatory sub-units of protein phosphatases, is a major determinant of the sensitivity of arabidopsis plants to salinity (Christophe and Michel, 2000).
During this period, the accumulation of Na^{+} in certain halophytes occurs only in the roots and in the hypocotyl, with exclusion in this case being ensured by the aerial part (Levitt, 1980).
In glycophytes, such as beans, plants generally exclude sodium from their leaves, although this ion is often accumulated in the stems and roots (Hubac, 1990).

- **accumulation**

ions can generally accumulate preferentially in the cells or specialised tissues of the root, stem or leaf, as in sorghum exposed to Na Cl, which concentrates Cl^{-} ions in the parenchymal cells of the leaf sheath (Boursier and Lauchli, 1989).
In Atriplex halimus they are sequestered in the organs.

specialised hairs such as vesicular hairs to avoid the toxic effect of salt at the level of parenchymal cells and conductive tissues (Levitt,

1980).According to the same author, in salt glands, Na^{+} is generally transported along an electrochemical gradient: it is not pumped but is electrically coupled with Cl^{-}.Accumulation occurs more in the leaves than in the stems and large roots; at the cellular level, it occurs in the vacuole. It can occur in large vacuoles, resulting in succulence of the plant (Leclerc, 1999).Accumulation is continuous, since water uptake is continuous at root level, and blocking salts is hardly possible, as it would be too costly in terms of active transport (Gerhard, 1993).
Salt can also be actively accumulated in the hairs, which then die, shed and are replaced by new ones (Leclerc, 1999).

- **dilution**

In a salty environment, growth is closely dependent on the maintenance of cell turgidity.
The dilution of absorbed salts is very often linked, in resistant plants, to high water retention and the development of succulence (Hamza, 1980).
In many halophytes found in salt marshes, large quantities of salts can accumulate in the leafy stems without resulting in a high concentration of salts, because as the salts reach the aerial organs, these become fleshy through the development of large water-rich cells that dilute the salts (Levitt, 1980).

b-Tolerance to low water potential

Tolerance is the set of mechanisms released by the plant in response to an imposed constraint, enabling it to get through this critical period and reproduce. This resistance can be acquired in two different ways. It can be achieved at different water potentials; high or low (Hisiao and Acevedo, 1974 in Rekika, 1997).

- **osmotic adjustment**

In the case of a drop in water potential, tolerance is expressed by the maintenance of turgidity, made possible by osmotic adjustment phenomena (accumulation of solutes) (Monneveux and This, 1997). Osmoregulation is an important adaptation mechanism for salinity-tolerant plants (Torrecilla, 1993; Azevedo et al., 2004).
Malate, the synthesis of which is carried out by PEPC, has frequently been considered as one of the organic solutes capable of lowering the potentialThis metabolite could play the role of osmoticum and thus participate in the mechanism of adaptation to salt stress. In fact, the

high activity of PEPC in the presence of Na Cl in a tolerant variety of wheat can be interpreted as a process that counterbalances the excessive penetration of inorganic cations, particularly Na+, by organic acids, mainly malic acid, which explains the improved salt tolerance (Whittingten and Smith, 1992).

In a salinity-stressed plant, osmotic balance is maintained by the accumulation of oses, polyols, amino acids (especially proline) and potassium salts.the balance between the water potential of the cytoplasm and that of the vacuole and apoplasm (Leclerc, 1999). These solutes help regulate pH variations in the cytosol and detoxify excess NH_4 (Gilbert et al., 1998).The accumulation of organic compounds in the roots favours the absorption of water from the soil (Robinson et al., 1997). In this case, the cells exert osmotic pressure, which creates a call for water and consequently lowers the water potential in the cells. This adjustment enables the growth rate and photosynthesis to be maintained (Azevedo et al., 2004).Da Silva et al., 2003, showed that the level of glycine betaine accumulation is correlated with the degree of salinity tolerance.In addition to its osmotic role, glycine betaine stabilises the structure of intrinsic PSП proteins, protects plasma membranes and activates enzymes in barley and rice (Melloni et al. , 2004).

According to the same author, salinity does not affect the accumulation of proline, which does not appear to play an important role in salinity tolerance mechanisms.Certain ions such as K^+ , Ca^{++} and Na^+ are also involved in osmotic adjustment. These ions need to be accumulated in the vacuole, as high concentrations are toxic to the cell. Transporting these ions to the vacuole requires energy (Ungar, 1996; Azevedo et al., 2004).

- **Tolerance to dehydration**

The superiority of a salinity-tolerant species is shown by greater resistance of tissues to dehydration and by a greater speed of absorption and transport of chlorine and of sodium in the leaves.

Numerous studies have shown the role of cell membranes in plant resistance to dehydration (Pha m thi et al., 1985).

In the case of certain species sensitive to dehydration, the general organisation of the cell may be affected, insofar as it leads to the destruction of certain cell organelles. Tonoplasts split into small vacuoles, mitochondrial cristae degrade and chloroplasts lose their molecular organisation (Da Silva, 1976 in Adjab, 2002).The degradation of membrane lipids and proteins also greatly disrupts cell function and causes a reduction in selective permeability, which influences intercellular molecular exchanges and electron transport (Nilson et al., 1996).Christophe and Michel in 2000 showed that a motif of five base pairs, known as DRE (dehydration responsive element) or LTRE (lowtemperature responsive element) plays a key role in gene induction by dehydration, cold and salinity independently of abscisic acid.

c- Tolerance to high water potential

- **reducing water loss**

The reduction in the flow of water through the plant shows, however, that the action of salt is similar to that of drought, even when the plants have adjusted their concentration.

According to Turner in 1986, the maintenance of a high water potential in the leaves is linked to the plant's ability to extract water from the soil and its capacity to limit water loss through transpiration.To minimise these losses, the leaves reduce their size and may even disappear; they diminish in size by curling up and folding in on themselves (Bidinger et al., 1989), so leaf fall in summer for plants in semi-arid regions considerably reduces transpiration (Heller et al., 1998). In addition, the hairiness and light colour of the leaves induce an increase in reflectivity which leads to a reduction in water losses (Al hakimi, 1993).Reducing the stomatal opening makes it possible to rapidly preserve the plant's water status, but is accompanied by a reduction in gas exchange and therefore photosynthesis (Levitt, 1980).

The operating principle of stomata is essentially based on variations in the turgor potential in the guard cell.are determined by numerous factors, some of which are linked to the environment and others to the plant itself (Heller et al., 1998).

On the other hand, the stomata establish a link between the external environment and the leaf parenchyma, with exchanges involving water and CO_2 . When the stomata are closed, the plant limits its transpiration but at the same time slows down the entry of CO_2 (Semiani, 2000).Other authors attribute the mechanism of stomatal closure to hormonal control (abscisic acid) (Tardieu et al., 1990); and also ionic accumulation (K^+ Cl^- and Na^+) (Alarchon et al., 1993).

Potassium plays an important role in the osmotic regulation of the

opening movements of tomatoes. Under the effect of light, the guard cells, probably by means of an ion pump, accumulate the active ions K^+, which are released at again at the next closing time. Essential energy is thought to be supplied by oxidative phosphorylation (Gerhard, 1993).In salinity-stressed maize, the reduction in potassium content in leaves and roots is due to competition between this ion and sodium, resulting in a metabolic change following ion toxicity at root level (Azevedo et al., 2004).
According to Christophe and Michel in 2000, there are potassium and calcium channels on the vacuolar membrane of the guard cell, which play a pivotal role in regulating cell volume.

- **Epidermal perspiration**

Transpiration occurs mainly through the stomata, but can also take place through the cuticle if it is sufficiently well developed.
This cuticular transpiration alone remains when the stomata are closed (Lafon et al., 1985).
In addition to stomatal and cuticular transpiration, young trunks undergo lenticel transpiration, with water evaporating through the cork lenticels (Heller et al., 1998).
Epidermal transpiration is defined as the rate of transpiration during minimal stomatal opening: It is the sum of residual cuticular and stomatal transpiration (Clark et al., 1991; Michow et al., 1989 in Rekika, 1997).The ratio of cuticular transpiration to stomatal transpiration (under favourable conditions) depends on the species. In species adapted to dry climates, it represents 1/50 of total transpiration (Heller et al. , 1998).

- **The root system**

The root system performs three essential functions in plants: it specialises in absorbing water and mineral salts from the soil, it is responsible for attaching the plant solidly to substrates by growing and branching downwards, and finally it accumulates reserves (Camfort, 1980).Callot et al (1982) in Guettouche 1998, point out that there is a close dependence between the aerial parts and the root system of plants. The root system receives the carbohydrates necessary for growth from the aerial organs, while supplying the plant with water and mineral elements. However, the relationship between the growth of the above-ground parts and that of the roots can vary greatly, depending on the conditions of the environment and the phenological

stage of the plant.It should be noted, however, that the water of imbibition which evaporates through transpiration must be permanently replaced by water from the soil, which is absorbed by the roots and maintains the functioning of the plant (Vilain, 1987).
According to Ali Dib et al, 1992; Sehabi, 2000, the maintenance of this type of functioning is ensured by the increase and depth of the main axis, enabling the water reserves of the deepest horizons to be exploited.A large surface area of root system development in barley can therefore make a major contribution to high yields under stress conditions (Gan et al., 2000).

II - THE PLANT

1-Biology of the species

Atriplex halimus

Plants in the Atriplex genus are found in most regions of the world. They belong to the Chenopodiaceae family and are characterised by their great diversity (Kinet et al., 1998).Atriplex halimus occurs spontaneously within a relatively vast area encompassing the countries of North Africa and the Near and Middle East, from the Canary Islands to Iran, via southern Algeria. In Europe, Atriplex halimus is found all along the northern shore of the Mediterranean and also in Bulgaria (Aouissset, 1992).Natural regeneration occurs mainly in the more or less saline depressions on the periphery of the sebkhas in arid regions, and is sometimes found on saline gypsum geological outcrops (Aouissset, 1992).Atriplex appears to be the most suitable plant for developing saline areas, since it can reduce soil salinity by around 1100Kg Na Cl/ha/year (Franclet et le Houerou, 1971).

Morphological appearance :

It is a shrub that grows in very dense clumps with a silvery appearance. When not subject to grazing, it can reach up to four metres in height and form forms that are difficult to penetrate (Negre, 1961).
The leaves of Atriplex halimus are two to six cm long, alternate, simple, entire, with a short stalk, rounded oval when young, triangular, more or less lanceolate later,silvery green and more or less fleshy, shiny, covered with blister hairs (Duperat, 1997).
The roots of this plant are large, initially spreading out obliquely and

then burrowing to a depth that varies with the soil and the age of the plant (Negre, 1961).
The inflorescence appears to be a raceme of cymes, comprising male flowers with five petals and five stamens and female flowers with no perianth, the gynoecium of which consists of an ovary surmounted by two styles, and enveloped by two opposite, triangular bracts (HCDS. 1996).According to Kinet et al. 1998, there is a temporal gradient in sex expression. Protandry seems to characterise the species.

Anatomical aspect :

The leaf: it is known that resistance to salt stress is linked to the plant's capacity to accumulate salt in its organs, and so it has been observed that the leaves of halophytes, particularly those of certain atriplexes, have salt glands that serve to actively eliminate and deposit NaCl in the large sphere-shaped hair cells that can be seen on the leaf surface (Guignard et al., 2004).Another anatomical feature of these plants is the presence of a perivascular sheath around the conductive tissues within the mesophyll (Heller et al., 1998; Elena et al., 2001). This sheath contains chloroplasts that are very poor in grana; the mesophyll, whose chloroplasts have a classic grana structure, has few or no lacunae (Martin et al., 1995). Communication between the cells of the mesophyll and the sheath is facilitated by the presence of numerous plasmodesmata. This structure enables these plants to develop a high level of assimilation.C4 photosynthesis in order to reduce water losses through transpiration in parallel with CO uptake$_2$ (Luttge et al., 1996).
The stem :

A cross-section of an Atriplex halimus stem shows the following: the primary woody liberal bundles, located in contact with the pith, have acquired secondary formations through the activity of the abnormally functioning cambium. It consists of a large meristematic zone producing conductive tissue on its inner surface. The derivatives differentiate either into islands of phloem superimposed on patches of vessels and parenchyma, or into fibres (Robert and Catesson, 1990). Later, supernumerary cambiums are successively formed. At the periphery of the section is a periderm formation which will differentiate xylem on the inside, while on the outside it has already differentiated a layer of suber (Gayral et al., 1961).

The root: in the root as in the stem, cambiums appear successively one after the other from the periderm to the pith, depending on the case, producing liber and supernumerary wood after the initial cambium has ceased to function (Gayral et al., 1961).

Ecological conditions

At the level of the whole plant, the halophytic character of the species has been confirmed (Kinet et al., 1998). This species is spontaneous in semi-arid to pre-Saharan bioclimatic zones, and its climatic requirements are as followsIt tolerates minimum temperatures of 5 to 10°C and maximum temperatures of between 30 and 30.8°C (HCDS, 1996).It requires an average annual rainfall of 150 to 1000 mm and is resistant to salinity and cold, even at absolute minimums of around -10°C (El Aich, 1987).It is a halophilous species with few soil requirements. It occupies sebkhas and the banks of wadis; it adapts well to deep, compact marl and silty soils. It can also grow on alkaline gypsum crusts from 0 to 1500 m (HCDS, 1996).

2-Species interest :

Animal feed - this species is a popular fodder for camelids, particularly sheep and goats, especially during periods of drought (Ben Ahmed et al., 1996; Mohd'elkhair et al., 2003). The fodder value of this species has been estimated at 0.56UF/Kg of matter.Its crude protein content varies from 12 to 18 kg of dry matter (Mohd'elkhair et al. , 2003).Reducing soil erosion and degradation - Installing a canopy of atriplex on otherwise bare saline sites goes a long way towards reducing wind and water erosion. This plant has a creeping growth habit and naturally picks its stems, two excellent characteristics for preventing soil erosion. In addition, the marcotting stems are better able to withstand intensive grazing (Pourrat et al., 1994).

Aesthetic improvement - Around many towns in arid areas there are large expanses of unattractive saline soil, which are a source of dust blown into the town by the wind. Atriplex bushes, which grow to around 1.50 to 2 m high, are a possible way of avoiding the dust and improving both living conditions and the aesthetics of the area (Le Houérou, 2000).

III- Characteristics water

Water is an essential element for all living things (Heller et al., 1998). It is always the essential weight component of an active cell. In general, the water content of plant tissue remains between 60 and 90% of fresh weight (Kramer et al., 1995). At the cellular level, water helps to maintain structures and enables metabolism to take place through the turgidity pressure it exerts on the walls. It also contributes to plant growth, controls various organ and cell movements and helps to elongate cells. In the body, it serves as a vehicle for nutrients (Heller et al. , 1998).

1- Water in the soil :

Water connections :

Water is bound to the soil constituents by two categories of forces: **a- osmotic forces:** these are due to the attractions exerted on the water by the ions in the soil solution.

b- **matrix forces**: reflect the bonds between the water and the

The figurative structure of the soil (matrix):

► **The forces of imbibition**: due to the electrostatic attractions exerted between the negative charges of the colloids and the positive poles of the water.

► **Capillary forces**: due to surface tension phenomena, which retain water in fine interstices. Like

osmotic forces, capillary forces, except in parched or saline soils, do not pose any problems for the plant and it is the greater or lesser colloid content of the soil that regulates, for equal humidity, the availability of water (Heller et al., 1998).

Agronomic aspects :

Field capacity :

When a soil is saturated with water (rain or irrigation), its relative humidity, once surface runoff has occurred, is its maximum retention capacity. Very quickly (a few hours), some of this water runs off by gravitation or evaporates, so the humidity is, if not constant, at least stable enough to provide an assessment of the water retention capacity. This moisture is the ch amp capacity (Heller et al., 1998).

Wilting point :

The permanent wilting point is the moisture content of a soil at which a plant irreversibly wilts (sand 5%, silt 12%, clay 27%, peat 50%) and is of interest because it is a soil characteristic; its value depends little on the plants (Morel, 1989).

The useful reserve

The difference between field capacity and wilting point measures the quantity of water available to the plant or useful reserve for the soil **R.U**(Clement, 1998).
Part of the **UC** is absorbed by the plant without reducing its growth: this fraction is called the readily usable reserve.
R.F.U and is most often estimated at two-thirds of the value of the

R.U in poor soil and between half and two-thirds in deep soils (Gate, 1995).

2- Water in plants :

The water content of plants

It is determined by comparing the mass of fresh matter with that of dry matter obtained by dissipation, and depends on the organ and species, under given conditions.
This content is a genotypic characteristic that is linked to the plant's ability to maintain a level of water in the leaf that is capable of guaranteeing the continuity of metabolic activity (Heller et al., 1998).

Water potential

The water potential (Ψ) expresses the binding state of water within plant tissues.Ψis a negative thermodynamic quantity (Heller et al.,1998) which determines the direction of water exchange between two points of
It is measured in units of pressure: equivalent to energy per volume (Laval-Martinet al., 1979).
The potential difference is the driving force behind all movement of water in the organism, determining its direction, i.e. towards the place where the potential is lowest (most negative), in the direction of a decreasing potential gradient (Clement et al., 1998).
In the vacuolated cell, the water potential is determined by the water potential of the cell juice, i.e. by the osmotic potential of the vacuolar contents (Gerhard, 1993).

According to Heller et al, 1998, osmotic forces are mainly involved in the vacuole, and the osmotic potential they create, due to mineral and organic ions, is always negative and of equal and opposite value to that of osmotic pressure.

Water absorption

Roots act as an interface between the plant and the soil, providing contact with the environment. The method and extent of soil colonisation will therefore largely determine water and mineral ion uptake (Hopkin, 2003).The root can only absorb if its suction potential is smaller than the soil potential (Leclerc, 1999).

The water potential of the cells decreases steadily in the bark from the absorbing hairs to the endodermis and, although it then rises slightly, there is still an appreciable difference between the potentials of the vessels and those of the soil (Lafon et al., 1996; Heller et al, 1998).Furthermore, the turgidity of parenchymal cells is generally higher than that of vessels, which means that water tends to penetrate the latter due to the mechanical pressure exerted by the walls of neighbouring cells (Heller et al., 1998). The diffusion of water towards the xylem tissues could not therefore take place without this intense active transport of ions, at the origin of what has been called root flushing, which only takes place if the roots are well aerated (Laval Martin et al., 1995).Another mode of regulation has recently been discovered:Major intrinsic proteins in biological membranes function as water channels, i.e. facilitate the diffusion of water, known as aquaporins (Heller et al., 1998; Christophe et al., 2000), and the specific expression of numerous aquaporin genes in plants suggests that these proteins are involved in water transport, in various organs and at various stages of plant development (Chistophe et al., 2000).

Water transit

Along its path through the bark tissue to the central cylinder, the water and the dissolved ions it contains take two routes. Firstly, it makes its way through the walls, diffusing freely in the interfibrillar spaces, and propagates relatively as far as the endodermis: this is known as apoplasmic transport (Lafon et al., 1996). However, deposits of hydrophobic substances make the cell walls of the endodermis impermeable, forcing the water to take the only possible route: the symplasmic route (Gerhard, 1993). The water absorbed by the roots is channelled towards the central cylinder, from where it is discharged into the vessels of the wood or xylem, containing mineral salts in solution, which are also absorbed from the soil. It then constitutes the raw sap (Heller et al., 1998). To a first approximation, this pathway is centripetal and perpendicular to the axis of the roots. It is therefore often referred to as **horizontal transit** (Lafon et al., 1996).

The horizontal pathway is followed by the essentially upward migration of raw sap. This involves vascular conduction or translocation in dead cells emptied of their protoplasm, which play a passive role and whose alignments constitute the vessels. These movements o f water end in transpiration (Binet et al., 1967).

Several mechanisms are involved to explain the ascent of raw sap in the stem. The first is capillarity (Heller et al., 1998), root growth: this is an active ascent that corresponds to real work by the cells of the root (Binet et al., 1967) and plays an important role when transpiration is low, but more often than not, it is transpiration that is the main driving force behind the ascent of sap (Heller et al., 1998).

The effect of salinity

Salt stress is defined as the presence of excessive concentrations of soluble salts in the soil, resulting in damage to the plant ranging from a slight drop in yield to total deterioration of the plant (Amrar, 1993; Wiliam et al., 1998). According to the same author, it is a highly complex phenomenon involving soil-water interaction in relation to a number of factors, all of which influence the extent of stress and the damage it causes.

Salinity affects all the plant's physiological processes. In particular, it reduces height growth. However, some wheat varieties tolerate slightly brackish water better than others (M'barek et al., 2001).

Growth inhibition may or may not be reversible. It depends on the level of concentration and the duration of exposure. This reduction in growth is attributed to the limited number of cells or their small size (Amrar, 1993).

The depressive effect of salt affects aerial organs more than roots, and is most apparent in young, growing leaves. This difference in sensitivity between the absorptive and photosynthetic organs is characteristic of glycophytes (Brugnoli and Björkman, 1992; Bernstein et al., 1993).

In 2001, M'barek et al. showed that the growth of stems, leaves and roots is significantly reduced when salinity exceeds 4 g/l in certain wheat varieties.Mozafar et al., 1984, inPourat et al., 1994 observed, on Atriplexhalimus grown on nutrient solutions with different Na+ and K^+ concentration ratios, that the best growth results wereobtained when the ratio was equal to 1. This would be due to a positive biochemical interaction between these two elements, perhaps linked to an ionic balance favourable to the functioning of certain enzymatic systems.Salinity also affects plant development. For example, NaCl delays flowering in rice cultivars (Amrar, 1993). In tomatoes, salinity reduces plant height, fruit size and favours earliness, but increases the senescence of bean leaves and the number of flowers per plant in cotton (Levitt, 1980).

CHAPTER II
EQUIPMENT AND METHODS

A- Plant material

The effect of salt stress was studied on Atriplexhalimus plants taken from the same mother plant at the senia university station (Oran). Atriplexhalimus seeds were manually stripped of their fruiting valves in order to obtain a better germination rate and rapid emergence. They were then disinfected by soaking in 8% sodium hypochlorite for 5 minutes and rinsing with distilled water.

First, the seeds are placed in peat-filled trays soaked in distilled water. As soon as the first leaves appear, the seedlings are transplanted individually into pots.

B-METHODS

1 - Experimental conditions a- The experimental site

The study was conducted in a greenhouse under semi-controlled conditions at the Institut des Sciences Agronomiques de Tiaret. During the experiment, the daytime temperature was maintained at around 28°C and the night-time temperature at 15°C, while relative humidity was approximately 70% and the degree of illumination was maintained at 12 hours by artificial light.

b- the substrate

The trial was conducted in plastic pots with an internal diameter of 15cm and a height of 18cm. Each pot was filled with a substrate consisting of a mixture of sand and peat with proportions of 2 -1 and a retention capacity estimated at 30%.

The sand used is first washed in the spirit of salt, rinsed thoroughly with distilled water to remove chlorides and carbonates and then air-dried.

c-the experimental set-up

Atriplex plants are treated with two types of salt

- The NaCl+CaCl solution$_2$ at 400 meq/l and 600 meq/l.
- Seawater diluted to 50% and undiluted (pure seawater).

Each treatment consisted of 15 replicates.

The substrate is irrigated at ch amp capacity with a daily watering dose based on the daily evapotranspiration of each pot. After the first three lcavcs appcar, the irrigation water is replaced by a commercial nutrient solution diluted to 1/1000 ème , until salt stress is applied.
With the plants corresponding to the stressed treatments, irrigation was applied every three days until the end of the experiment, using the Na Cl+CaCl solution2 and seawater.

Table 2 - Composition of the nutrient solution

	The element		at	g/ 1	Weight
	KNO 3			191.920	
	(NO3)2 Ca4 H2 O		%		
The major components	NO3 NH4 SO4 Mg7 H2 O PO4 H2 K			129.8 210 61.5	
	PO4 K2 H3 H2 O			54.4	
				34.23	
		Cl2 Mn4H2 O	1.8		
		CuSO4 5 H2 O	0.176		
		ZnSO4 7 H2 O	0.219		
The trace		H3 BO 3	2.861		
elements		Mo7 O24 (NH4) 7	0.285		
	H2 O		0.05		
		Ferric EDTA			
		C10 H12 Fe N2 Na O 8			

2 - Analysis techniques

All the tests were carried out on the two trials, with measurements taken separately on different leaf stages depending on the water parameters studied.

a- physiological parameters

Changes in the plant's water status were measured by monitoring the plant's water content and osmotic potential. To do this, two series of measurements were taken: at the start and at the end of the stress.

• The loss of water through the excised leaf (RWL) and the water content of the leaf (RWL). of water (TRE)Water loss is often considered to be a good indicator of the plant's water status, as it is linked to cell volume and transpiration rate (Schonfeld et al., 1988).characteristic is determined using the Ladiges (1975) method, used by Clarke et al., 1982 and Rascio, 1988.To measure it, the leaf is cut at the base of the leaf blade and immediately taken to the laboratory, where it is left on the bench under ambient laboratory conditions. Half an hour later (theoretical partial closure of the stomata), an initial weighing (P1) was carried out. After one hour of dehydration, a second weighing (P2) was carried out. On the same sample, the leaf surface area (LSA) was measured (in cm^2) using an electronic planimeter. Residual transpiration was calculated as follows:

RWL (mg H_2 O .cm-^{2}.m^{-1}) =P1 -P2 / SFxTx

RWL = perspiration rate.

P_1 = weight of leaves after one hour. P_2 = initial weight.
Sf= leaf area. Tx= time (60 min).

The relative water content (RWC) was measured on a well-developed leaf using the method of Barrs and Weatherley (196 2). The leaf blade, excised at its base, was immediately weighed to determine the fresh weight (Pf), and the cut end was then placed in a test tube containing distilled water and placed in the dark at 2°C for 24 hours. Weighing the leaves after rehydration gives a rehydration weight (Pr).
The dry weight (Ps) is determined after the samples are placed in an oven at 80°C for 48 hours. The TRE is calculated using the following equation:
ERR (%) = Pf-Ps / Pr-Ps x100

Pf= fresh weight,

Pr= saturation weight, Ps= dry weight

• **Osmotic potential**

This parameter is measured using an automatic osmometer (VAPRO) (fig.3), which determines the evaporation point of aqueous solutions. Lowering the freezing point compared with distilled water gives a direct measure of the molar concentration.

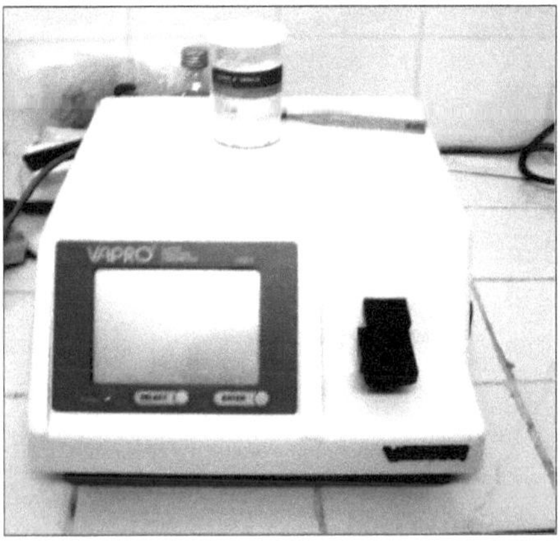

Fig.3 - Photo showing the automatic osmometer (VAPRO) for measuring osmotic potential.

- **stomatal resistance**

Stomatal resistance ($s.cm^{-1}$) is measured using a porometer (Fig.2). The sheet used for the measurement is placed in such a way that the thermocouple of the device is in contact with the middle of the sheet. Readings are corrected for variations in temperature. relative air humidity and leaf and ambient temperatures, in order to express the results in relation to 20°c.

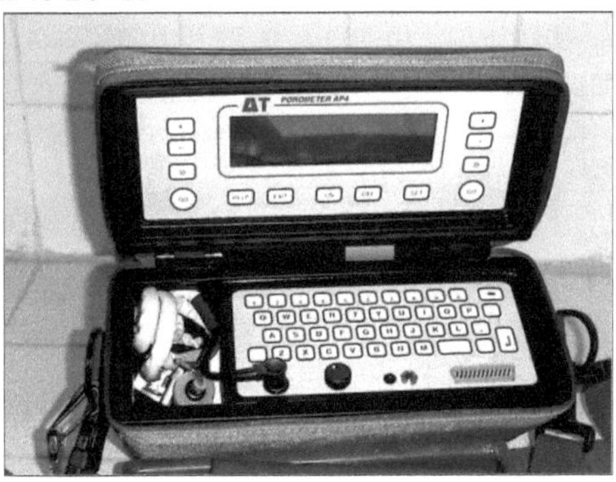

Fig.2 - Photo showing a porometer for measuring resistance and stomatal conductance

b - Biochemical parameters

Wax content :

The leaf blade of the penultimate leaf is excised at the base and then dipped into a test tube, washed and weighed beforehand (P1). Chloroform is added to the tube to rinse the leaf a n d extract all the wax. The leaf area (LA) was measured using an electronic planimeter. The test tube containing the chloroform and wax was dried in an oven for 24 h at a temperature of 45°C, then weighed (P2). The wax content per unit area (cm^2) is determined by the difference in the weight of the tube before and after wax extraction.
Wax rate ($mg.cm^{-2}$)= P2 -P1/SF TC= wax rate.
P_{TE} = Weight of the tube after wax extraction.

P_{TI} =Initial weight of the test tube. S_F = Leaf area.

Determination of soluble sugars

The oses (glucose, fructose and sucrose) are measured by the Shields and Burnet (1960) method used by Kourichi (1992) and Rekika (1997). These sugars are extracted after maceration in a solvent capable of solubilising them and blocking the enzymatic activity likely to degrade them. The principle of the reaction is based on the condensation of the degradation products of the neutral oses by sulphuric acid: the latter, which is highly concentrated, transforms the oses at high temperature into furfural derivatives which give a blue-green colour with anthrone, measured by colorimetry at 585 nm. Plant material taken (approximately 100mg) from the middle third of the leaf is left in 5.25 ml of 80% ethanol for 24 hours; 2 ml is taken of the extract previously diluted 10 times with 80% ethanol (extract A); the reagent (extra it B) is added to the extract. was prepared 4 h before the assay at a concentration of 2 g of pure anthrone added to 1000 ml of sulphuric acid (H2SO4). 2 ml of the extract (A) was then taken and added to all the test tubes, all of which were kept in melting ice. After shaking, the tubes are placed in a water bath at 92° C for a few minutes. 8 min; then cooled for 30 min. in the dark. The absorbance is read on a spectrometer at a wavelength of 585 min and the concentration is expressed in µmol / g MS

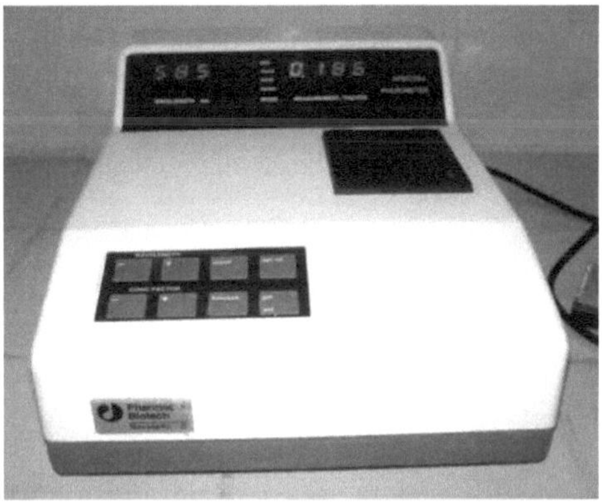

Fig.5 - Photograph of the spectrophotometer (VAPRO) used to measure absorbance.

c. Morphological parameters of the above-ground and underground parts

• the air part

The dry matter of the above-ground part was obtained by placing in an oven at 80°C for 48 hours. This measurement was used to establish the ratio of root part to aerial part

• The underground section

Once the aerial part has been studied, the pots are dug up and their contents collected in tanks filled with water. The roots are carefully washed in order to identify the following morphological characteristics:

- the length of the main root axis under both stresses

(to $NaCl+CaCl_2$ and seawater).

- The dry weight of the roots: this is obtained by oven drying.

at a temperature of 80°C for 48 hours.

d- anatomical parameters

The internal factors involved in limiting water loss are assessed by making anatomical sections of the leaves and roots. Tissue sites are assessed using the method of Martojaet al., 1967 used by Augier et al., 1982 . Several steps are required to make anatomical sections of the leaves:

• from leaves

Fixation is the phase that blocks all physiological activity in the plant fragment by hardening it. The material is placed in a test tube containing FAA (formole acetic acid) fixative and the samples are washed in running water for a period equal to the fixation time (24 hours).

Dehydration is carried out by successive passages through ethanol baths of increasing concentration (50, 90 and 100%), with the plant fragment losing the water in its cellular content.

Impregnation consists of immersing the sample in an equal mixture (50%, 50%) of toluene and paraffin. The first stage is carried out in an oven at 80°C for 4 hours. The second stage is carried out using only pure paraffin in order toto eliminate the toluene from the plant fragment, the whole is returned to Oven dried at 80°C overnight. Inclusion is the phase during which the pure paraffin blocks in which the samples are placed are produced. These blocks are attached to a motorised microtome of the LEICA RM 2145 type to produce sections with a thickness of 7 μm.

Once obtained, the sections are placed on slides spread with one or two drops of gelatinised water to ensure total adhesion of the sample to the slide.Dewaxing is carried out twice in a toluene bath: first hot to remove the paraffin and then 24 hours later to prevent the sections from becoming detached.Staining with a dye based on carmine alumina and methyl green, in which the sections are soaked for 10 minutes. They are then removed and washed under running water. One or two drops of Canada balsam dissolved in toluene is spread on the slide, enhancing optical transparency, preserving the section and intensifying staining.

Measurements are taken using a microscope equipped with a graduated eyepiece micrometer. The values obtained are multiplied by the micrometric coefficient specific to each magnification in order to convert the results into micrometres (μm).Measurements were taken on the main vein using a microscope equipped with a graduated eyepiece micrometer:

- ❖compaction of the mesophyll,
- ❖the size of the mesophyll (ventral side and lateral side)back),
- ❖dimensions of the perivascular sheath cells,
- ❖the diameter of the metaxilimus vessels.

• from the roots

The histological technique used was the free-hand sectioning technique. This was carried out on FAA-fixed Atriplex roots in the following stages: the roots were sectioned to obtain cross-sections using a scalpel blade. The cross-sections must be thin and parallel to the cutting plane.
The sections were stained using Carmine alumina and methyl green.
The whole is fixed with a slide and the measurements based on the observations are as follows:

❖The diameter of the metaxylimeric vessels.

❖The number of cribrovascular bundles.

❖The thickness of the cortical parenchyma.

❖The length and the width of cells of the parenchyma cortical.

❖the thickness of the rhizodermis.

Fig.4 - Photo of the microtome (LEICA 2145)

CHAPTER III

WATER CHARACTERISTICS A - PHYSIOLOGICAL PARAMETERS

1 - relative water content

Leaf water content provides information on the relative turgidity of tissues and is one of the criteria used to assess stress tolerance. It is linked to the plant's ability to maintain a level of tissue hydration that guarantees the continuity of metabolic activity. These results have been established and confirmed in numerous studies, including those of (Monneveux et al., 1992). To assess the water status of plants tested under salt stress, we monitored variations in relative water content in the first phase, one week after application of the salt solution, and in the second phase, after four weeks.

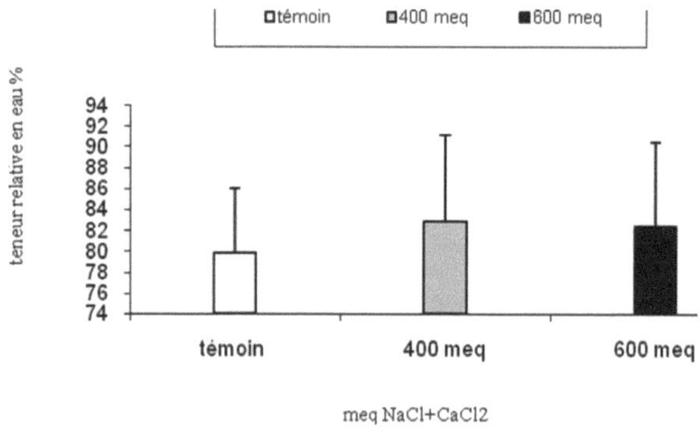

Fig.6 - Moisture content (%) of leaves of 120-day-old Atriplexhalimus L. plants, measured after one week of **NaCl+CaCl2** stress.

a- relative water content 1 (RWC1)

- **after a week of salt stress**

- to $NaCl+CaCl_2$

The results show (Table 3) that the variations observed are independent of the salinity levels imposed ($p>0.05$).Figure 6 shows that the relative water content varies with increasing salinity. When the plants are watered with nutrient solution alone, the water content reaches 79.83%. Under saline treatment, this content is maintained at a sufficiently high level (82.82% and 82.45%).

-seawater .

A study of the statistical results obtained by estimating the RWC (table 3) shows that seawater irrigation is a source of variations in the levels of this parameter ($p<0.05$). Thus the reduction in the salt content of the substrate is accompanied by a consequent increase in the water content. This is confirmed by the clear relationship between the salinity factor and the expression of this parameter (r= -0.661**).

The response of plants watered with seawater does not seem to be expressed in the same way as that described above for the NaCl+CaCl treatment$_2$. Nevertheless, Figure 7 shows that the relative water content varies from 89.05% in the control plants to 79.17% in the plants treated with 50% seawater; this content does not vary much in the plants (78.92%) when they receive undiluted seawater.

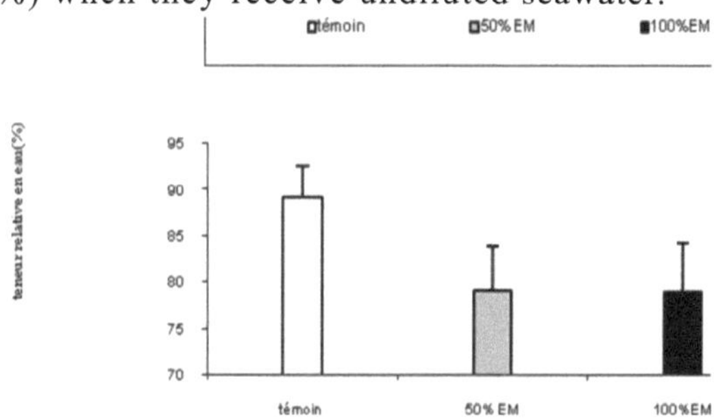

Fig.7-Moisture content (%) of leaves of 120-day-old Atriplexhalimus L. plants, measured after one week of **seawater** stress.

• after four weeks of salt stress (RWC 4)

-to $NaCl+CaCl_2$

The relative water content is estimated at a second measurement stage in order to elucidate changes in the water status of plants that remain under stress for longer. The answers we are looking for are mainly based on the intensification of the effect of the stress over time and the time needed for the resistance mechanisms to emerge.The second period is characterised by the fact that the intensification of the saline treatment is accompanied by an increase in the level of hydration. Figure 8 and Table 3 show that the highest water content was observed in plants treated with 600 meq (76.02%) and 400 meq (75.21%). In the absence of salts, the leaves had a water content of 72.19%.The same pattern of changes in relative water content between the different treatments is observed between the measurement periods. This indicates that the development of plants in a saline environment has not no influence on this characteristic.

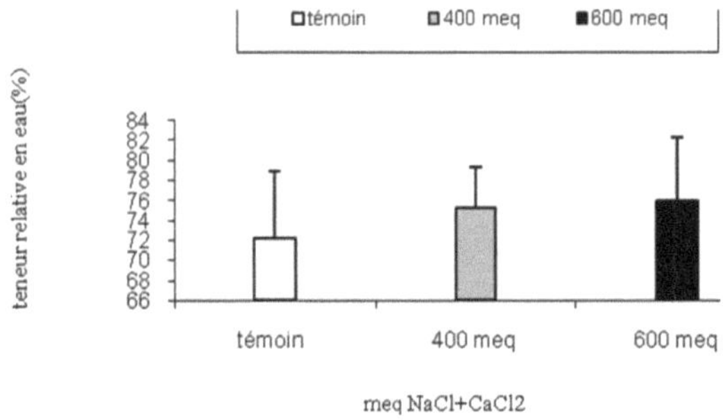

Fig.8 - Moisture content (%) of leaves of 120-day-old Atriplexhalimus L. plants, measured after four weeks of **NaCl+CaCl 2** stress.

- **with seawater**

The analysis of variance (Table 4) showed that the development of this characteristic was strongly influenced by the salt regimes adopted ($p<0.05$), with an increase in salt concentration leading to a reduction in leaf tissue water content ($r=-0.367$).Figure 9 shows that, as the relative water content decreases with salinity, plants growing in a saline environment have a lower water content than those grown without salts. It is also shown that, in the control treatment, the water content appears to be maintained at a high and stable level (86.27), while in the treatment carried out at 50%, the TRE falls more or less significantly to 75.29%, and rises sharply (80.16%) when the plants are watered with undiluted seawater.

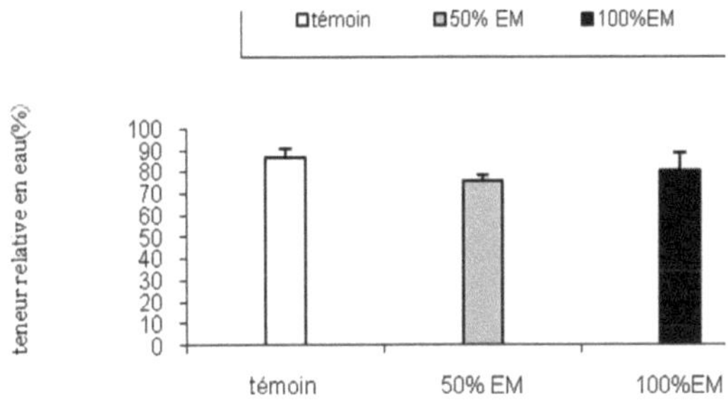

Sea water

Fig.9 - Water content (%) of leaves o f 120-day-old Atriplexhalimus L. plants, measured after four weeks of **seawater** stress.

2 - Perspiration rate

-to $NaCl+CaCl_2$

Perspiration measurements are taken from the 4th week of salinity application.Statistical analysis of the results obtained (Table 3) shows that the variations observed are independent of salinity levels ($p>0.05$). This indicates that fluctuations in the concentration of salts in irrigation water have only a slight influence on variations in the intensity of cuticular transpiration, depending on the method used.The results obtained by estimating this characteristic (fig. 10) show that the highest transpiration is obtained for plants grown without salt ($0.019 mg.cm^{-2}$ .mn). This transpiration decreases sharply in plants fed with 400 meq of Na Cl $+CaCl_2$ ($0.007 mg.cm^{-2}$ .mn), then increases when the plants are irrigated with 600 meq of Na Cl $+CaCl_2$ (0.014 $mg.cm^{-2}$ .mn).According to the data collected, the absence of any differential regulation of transpiration indicates that the absence of any externalization of any mechanism involved in the limitation of water loss.

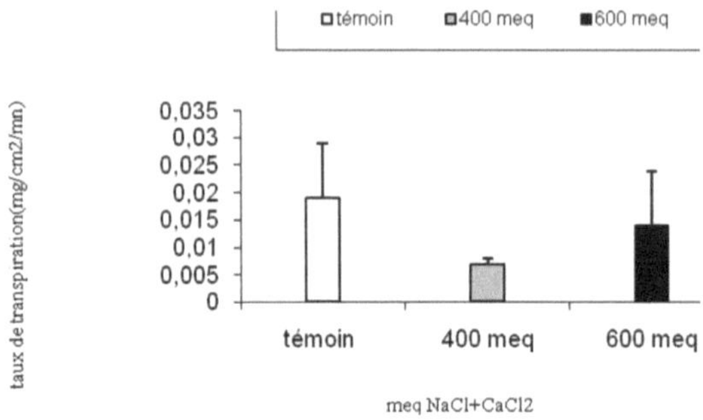

Fig. 10 - Leaf transpiration (mg.$^{cm-2}$. min) measured after 4 weeks of **NaCl+CaCl 2** stress in 120-day-old Atriplexhalimus L. plants.

- with seawater

On the other hand, irrigation with seawater at different concentrations resulted in significant fluctuations in the values of this parameter ($p<0.05$), so that water loss through the leaves was greater in the control plants than in the plants subjected to salinity.Figure 11 and Table 4 show the variations in the amount of water lost per plant. This figure shows that the control plants lost the greatest quantity, with 0.047 mg.cm-2 .mn.however, this quantity is much lower in plants watered with seawater than in plants receiving only 50% seawater (0.009mg.cm-2 .mn compared with 0.018 mg.cm -2 .mn).This shows that the increase in salinity is accompanied by a decrease in the transpiration rate for all the plants tested, contrary to what was observed in the previous study.For the first device, this is justified by the strong relationship between the two variables ($r2 =-0.740^{**}$).

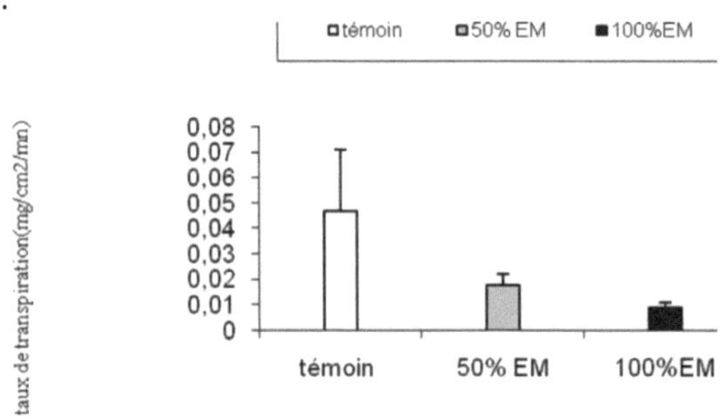

Sea water

Fig.11 - Leaf transpiration (mg.$^{cm-2}$. mn) measured after 4 weeks of **seawater** stress in 120-day-old Atriplexhalimus L. plants.

3 - Osmotic potential

• after a week of salt stress

-to $NaCl+CaCl_2$

Analysis of the data obtained (Table 3) shows that the expression of this characteristic is highly influenced by salinity ($p<0.05$). This indicates that salinity is a source of variation in the magnitude levels of this parameter.
The average results recorded (Table 3) show that the osmotic potential in the control plants reached -6511 Kp. In the 400 meq treatment, however, the difference was not very marked, with a value of -6857 Kp, while the osmotic potential remained lower in the 600 meq treatment (-9390 Kp).

• after four weeks of salt stress (PO4)

- to $NaCl+CaCl_2$

Estimation of osmotic potential with increasing duration of stress shows a clear reduction in its values.A study of the results reveals that the influence of salinity on the expression of this characteristic is weak ($p>0.05$).The results obtained following the estimation of the latter (Table 3) show that the highest value of osmotic potential is recorded in the control treatment, evaluated at -4552 Kp; this value decreased significantly in the two treatments subjected to salinity where this parameter reached - 5606 Kp and -9605 Kp respectively for the treatments carried out at 400 and 600 meq.

Table 3 - Statistical test of significance at P = 1% for physiological parameters of Atriplexhalimus L. plants stressed with **NaCl+CaCl2**

Variables	Test F	probability	Mean and standard deviation		
			Indicator	400meq	600meq
RWC1 (%)	0.643	0.532	79.83 ± 6.18	82.82±8.40	82.45 ±7.99
RWC4 (%)	1.655	0.205	72.19 ± 6.74	75.21 ±4.14	76.02 ±6.21
PO1 (Kp)	5.033	0.024	-6511 ±682.5	-6857 ±173	-9390 ±1752
PO4 (Kp)	13.42	0.069	-4552 ±1102	-5606 ± 296	-9605
T.t (mg.cm-2 .mn)	2.240	0.126	0.019 ± 0.01	0.007± 0.001	0.014 ±0.01
R.s (s.cm)-1	3.491	0.041	16.98 ±7.84	∞	∞

RWC1 (Relative Water Content) and PO1 (osmotic potential) = measurements taken one week after adding the NaCl+CaClsaline solution. 2

RWC4 and **PO4** = measurements taken 4 weeks after adding the NaCl+CaCl saline solution 2

T.t = transpiration transpiration rate and **Rs** = stomatal resistance

A study of the relationships established between the two variables reveals that the osmotic potential decreases with increasing salinity. This is confirmed by the clear negative relationship established between salinity and osmotic potential measured after 4 weeks of stress (r^2 =-894).*Overall, the same trends can be seen when comparing the two measurement periods, with in particular a reduction in water potential under the saline regime.

4 - Stomatal resistance

This parameter was also determined after 4 weeks of salt stress.

- to $NaCl+CaCl_2$

Analysis of variance (Table 3) revealed that variations in this parameter were highly dependent on fluctuations in salinity levels ($p<0.05$), so it appears that the increase in stomatal resistance is favoured by an increase in the salt content of the medium (r=0.379*).

The average results show that plants grown under saline conditions reacted by increasing their stomatal resistance, with values tending towards infinity for the two saline treatments (at 400 and 600 meq) and

16.98s.cm^{-1} for the control treatment.

-seawater

Analysis of the results obtained (Table 4) shows that the evolution of stomatal resistance follows the same trends observed in plants watered with the NaCl+CaCl solution$_2$, indicating the significant influence of salt regimes on the expression of this characteristic ($p < 0.05$).

The average results obtained (Table 4) show that stomatal resistance is higher in the treatment carried out without salts and that carried out with 100% seawater (∞), whereas in the treatment carried out with 50% seawater, resistance shows a low value estimated at 20.94 s.cm .$^{-1}$

Table 4 - Statistical test of significance at P =1% for physiological parameters of Atriplexhalimus L. plants stressed by **sea water**.

Variables	Test F	probability	Mean and standard deviation		
			witness	50% EM	100% EM
RWC1 (%)	7.912	0.006	89.05 ±3.45	79.17 ±4.77	78.92 ±5.33
RWC4 (%)	4.630	0.032	86.27 ±4.35	75.29 ±3.10	80.16 ±8.33
T.t (mg.cm-2 .mn)	9.170	0.004	0.047 ±0.024	0.018 ±0.004	0.009 ±0.002
R.s s.cm	2.800	0.100	∞	20.94±4.96	∞

B- BIOCHEMICAL PARAMETERS

1- Soluble sugars

- **at the start of salt stress**

-seawater

A study of the results (Table 6) shows that variations in salinity levels have only a slight influence on the fluctuations observed in the expression of this parameter ($p>0.05$). A weak positive relationship ($r^2 =0.189$) links the sugar rate to the increase in substrate salinity.

The average results show variability across the different treatments. In the treatment carried out without salt, the plants recorded the lowest level, estimated at 9 μM.g^{-1} MS. On the other hand, plants subjected to 50% seawater plants recorded the highest level of 12.5μM.g^{-1} MS, this value fell slightly in the 100% seawater treatment to 11.5μM.g^{-1} MS, but is still higher than in the control plants.

- **after four weeks of salt stress**

-to NaCl+$CaCl_2$

A study of the results obtained (Table 5) shows that the imposed salinity regimes consequently cause low-intensity fluctuations ($p>0.05$), indicating that variations in salinity levels are accompanied by slight fluctuations in soluble sugar levels.Although the influence is small, it is part of a negative trend, so any reduction in substrate salt levels is accompanied by an increase in leaf sugar content($r=-0.530$).The average results obtained show that the sugar levels in the leaves are very similar in the 400 and 600 meq treatments. They were evaluated at 16.2 and 16.1μM.g^{-1} DM respectively. The highest sugar levels were found in the control plants, with an average of 16.8μM.g^{-1} MS.

-seawater

In order to carry out a comparative analysis of the evolution of the accumulation of sugars in the leaves under the effect of the duration and intensity of the stress, a second dosage was carried out. At this stage of the analysis, the same results were obtained. They indicate that the influence of the salinity level on this parameter is weak ($p>0.1$). However, there was a positive relationship between the two characteristics ($r^2=0.487$).The results show that plants subjected to salinity reacted by increasing the quantities of sugars in their leaves. In fact, 34.5μM.g^{-1} MS was recorded in the 50% seawater treatment and 36μM.g^{-1} MS in the 100% seawater treatment. On the other hand, in the absence of salinity, the plants had an estimated level of 27μM.g^{-1} MS. Although the behaviour of the plants appears identical for the three treatments, during the two measurement phases, the increase in the degree of salinity led to a considerable change in the sugar content in the leaves.

2 - Wax content

-to NaCl+$CaCl_2$

The statistical study of the results obtained (table 5) from the evaluation of this characteristic shows that its development is independent of variations in salinity ($p>0.05$). A weak positive relationship is noted between the increase in salinity and the wax content (r=0.155).

Table 5 - Statistical test for significance of biochemical parameters of Atriplexhalimus L. plants stressed with **NaCl+CaCl 2** .

variables	Test F	probability	Mean and standard deviation		
			witness	400 meq	600 meq
S1 (μM.g-1 of PS) T. C (mg.cm-)2	2.274 0.637	0.145 0.541	16.8 0.28±0.26	16.2 0.50±0.54	16.1 0.41±0.20

S 1 = sugar content analysed in leaves one week (S 1) after stress.

T. C=t aux of wax analysed in the leaves à 4 weeks from stress.

On reading the results (fig. 12), however, we note variations in the magnitude of this parameter. Deposition was highest in plants exposed to 400 meq of salt, where 0.50 mg.cm-2 was recorded. In plants exposed to 600 meq of salt, the rate of deposition was lower.wax is estimated at 0.41 mg.cm-2 whereas in the control plants, the wax content decreases to a minimum of 0.28 mg.cm .$^{-2}$

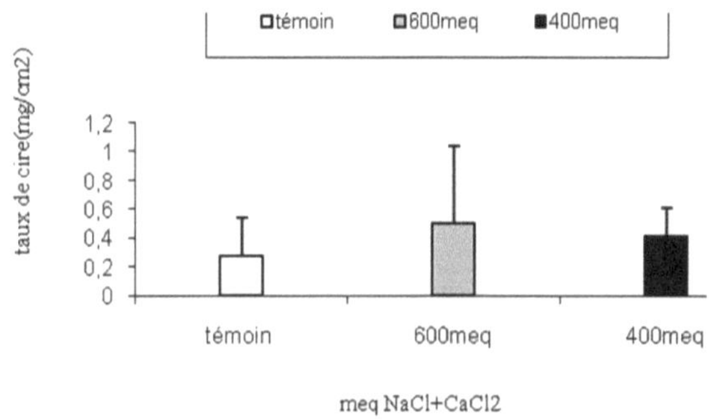

Fig.12- Wax content analysed in the leaves of 120-day-old Atriplexhalimus L. plants after 4 weeks of **NaCl+CaCl 2** stress.

-seawater

Analysis of the results obtained (Table 6) shows that the salinity levels imposed had only a slight effect on wax production (p>0.05).

Table 6 - Statistical tests Statistical test of parameter significance biochemistry of Atriplexhalimus L. plants stressed **by seawater**.

variables	Test F	probability	Mean and standard deviation		
			Indicator	50% EM	100% EM
S1(μM.g-1 of MS	0.398	0.680	9	12.5	11.5
S 4(μM.g-1 of MS	0.027	0.974	27	34.5	36
TC (mg.cm-)2	0.093	0.912	0.43±0.29	0.37±0.25	0.45±0.29

S1 and S 4 = sugar levels analysed in the leaves after one week (S1) and 4 weeks (S 4) of stress.

The average results obtained show that the wax content reaches 0.43mg.cm^{-2} under optimal conditions and under treatment with 50% seawater, the plants show a wax content of 0.2mg.cm. of 0.37mg.cm^{-2}, a level that is rather lower than that of the control, whereas in plants subjected to undiluted seawater, the wax level increased to 0.45mg.cm.$^{-2}$ To this end, it can be clearly seen that the wax content is not not affected by salinity. (r^2 =0.026).

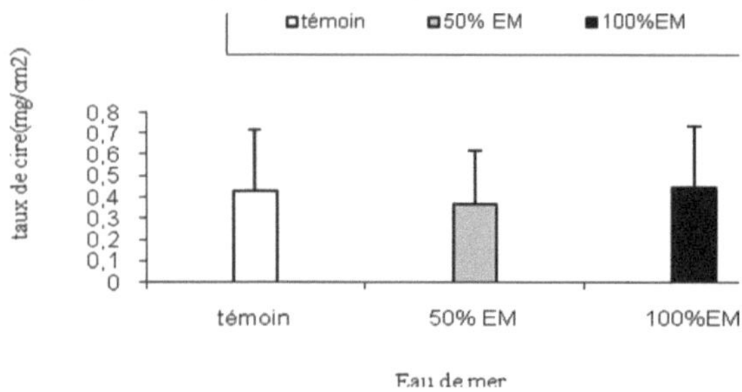

Fig.13 - Wax content analysed in the leaves after 4 weeks of **seawater** stress, plants from Atriplexhalimus L. aged 120 days .

C - MORPHOLOGICAL PARAMETERS OF THE ROOT SYSTEM

1- the dry weight of the roots :

-to NaCl+CaCl$_2$

Root dry weight is a morphological characteristic that is influenced by the salt regime applied ($p<0.05$), indicating that the concentration of salts in the irrigation water causes a reduction in the production of dry matter in the root system.

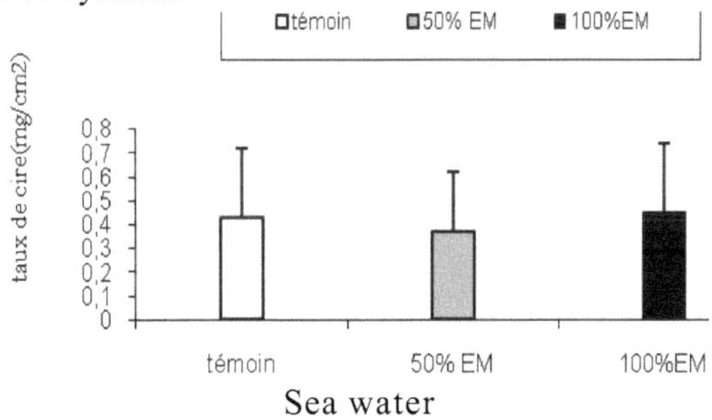

Fig.14 -Dry weight of roots estimated after 4 weeks of **NaCl** stress **+CaCl 2** , of 120-day-old Atriplexhalimus L. plants.

Analysis of the results obtained (Fig. 14) reveals that the plants tend to accumulate much more dry matter in their roots in the absence of salts, with an estimated value of 1.74 mg, whereas they do not retain this characteristic under saline conditions. The values recorded were virtually identical, at 0.49mg and 0.50mg for the treatments carried out at 400 and 600 meq respectively. Furthermore, the decrease in root dry matter production is accompanied b y an increase in the degree of salinity. (r=-0.781**).

- **with seawater**

Table 8 shows that the development of this parameter is highly dependent on variations in the study factor ($p<0.05$), which suggests that the saline regimes adopted allow significant fluctuations in this characteristic to occur.There was also a strong negative relationship between the two variables (r=-0.738 **).Figure 11 shows that the difference in the evolution of the dry matter production between control and control plants irrigated with 50% diluted seawater was not too marked: 1.74 mg and 1.54 mg respectively for the control treatment and the treatment with 50% seawater. This difference was clearly significant in plants irrigated with pure seawater, with 0.79 mg.

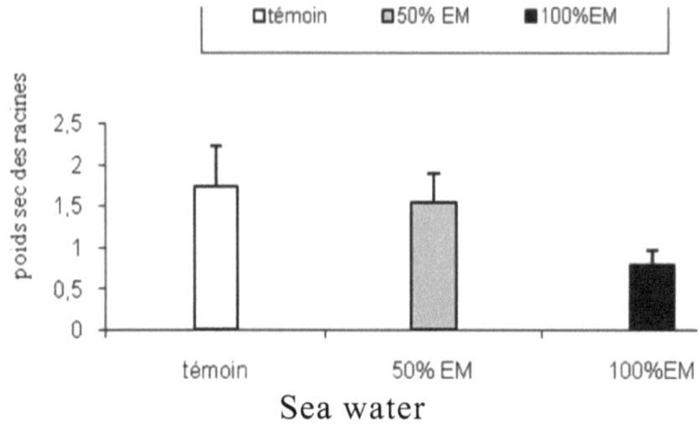

Fig.15 - Estimated root dry weight after 4 weeks of **seawater** stress in 120-day-old Atriplexhalimus L. plants.

2- The PSR/PSA ratio

This measurement was carried out in order to analyse the relationship between the growth of the above-ground parts (PSR, root dry weight) and that of the below-ground parts (PSA, dry weight of the above-ground part). It presents a highly debated characteristic.

- to NaCl+CaCl₂

Analysis of the results obtained (table 7) reveals that the variations observed in the ratio are independent of the degree of salinity, and this observation is confirmed by the absence of any significant effect ($p>0.1$), which seems to indicate that irrigation with salt water has only a slight effect on the development of this parameter.

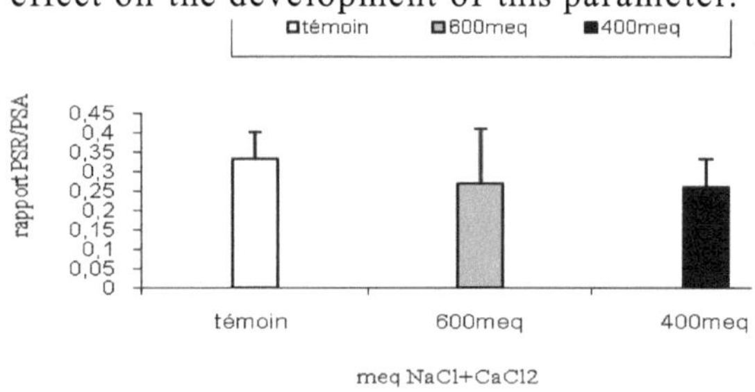

Fig.16 - PSR/PSA ratio after 4 weeks of **NaCl** stress **+CaCl 2** from 120-day-old Atriplex halimus L. plants.

Table 7- Statistical test of significance at P = 1% for morphological parameters of the root system of Atriplexhalimus L. plants stressed with **Na Cl+Ca Cl 2**

variables	Test F	probability	Mean and standard deviation		
			witness	400 meq	600 meq
P S R(mg)	26.57	0.000	1.74 ±0.49	0.49 ±0.34	0.50 ±0.16
PSR/ PSA	0.942	0.412	0.33 ± 0.07	0.27 ± 0.14	0.26 ± 0.07
Length(cm)	6.387	0.010	53.2 ±7.59	45 ±5.47	59 ±6.38

Figure 16 shows that this ratio reached a maximum value of 0.33 in the salt-free treatment, whereas in the salinity-treated plants, the ratio decreased and stabilised at around 0.27 in the 400meq treatment and at 0.26 in the 600meq salinity treatment.So, to sum up, the reduction in rat io is the higher the salinity level (r= -0.312).

-seawater

Looking at the average results (fig17), it can be seen that in the control plants, the ratio values recorded are the most High with 0.33, however, this ratio decreases, at level of treatment carried out à 50% water from sea water where the plants can be seen at 0.27. Plants exposed to seawater have a higher ratio than plants exposed to water.is around 0.20.

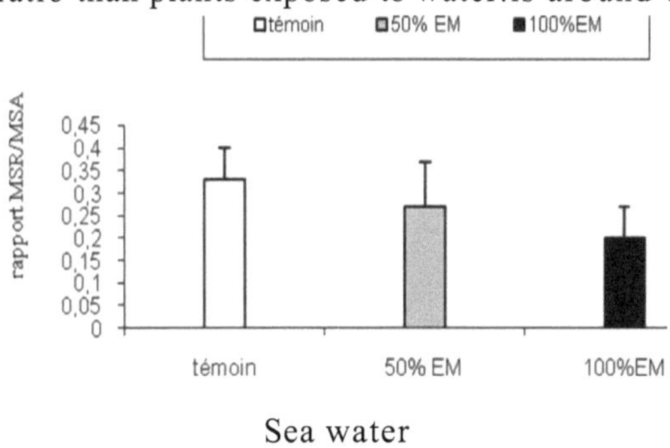

Fig.17 - PSR/PSA ratio after 4 weeks of **seawater** stress, 120-day-old Atriplexhalimus L. plants.

3- The length

Rooting depth can be measured by the length of the main axis.

- to $NaCl+CaCl_2$

In the light of the results, we note that the development of the main axis shows a strong variation in magnitude linked to the salt content of the substrate ($p<0.05$).

Measurements taken on the roots show that, according to Figure 18, the highest main root depth was observed in the 600 meq treatment at 59 cm, whereas rooting was shallow in the plants grown without salts at 53.2 cm. In contrast, the lowest rooting depth was recorded under the 400 meq treatment at 45 cm.

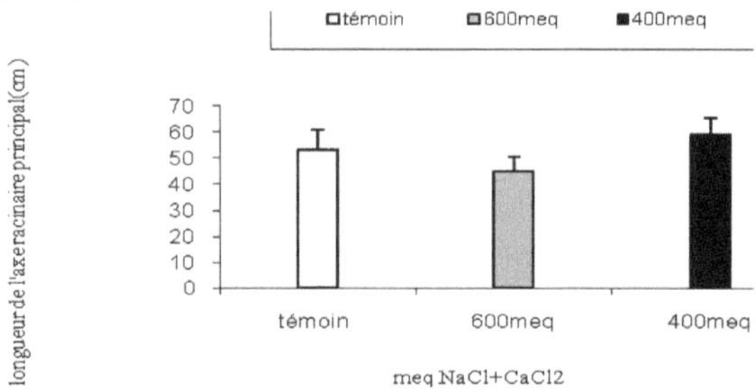

Fig.18 - The length of the main axis after 4 weeks of stress at the **NaCl+CaCl 2**, of 120-day-old Atriplexhalimus L. plants.

An increase in salt content was accompanied by a significant reduction in root elongation ($r=-0.478$).

-seawater

The results show that the elongation of the main axis seems to develop indifferently to the salinity regime ($p>0.1$). Variations in salinity thresholds allowed only very slight fluctuations in this parameter.

Table 8- Statistical test of significance at P = 1% for morphological parameters of the root system of Atriplexhalimus L. plants exposed to **seawater.**

variables	Test F	probability	Mean and standard deviation		
			witness	50% EM	100% EM
P S R(mg)	9.248	0.004	1.74 ± 0.49	1.54 ± 0.36	0.79 ± 0.18
PSR /PSA	3.803	0.053	0.33 ± 0.07	0.27 ± 0.10	0.20 ± 0.07
Length(cm)	1.748	0.216	53.20 ±7.59	49.02 ±9.77	43.90 ±5.74

Figure 19 and Table 8 show that the brackish water treatment showed the greatest reduction in the main axis for all plants (43.90 cm). On a scale of treatment with 50% diluted seawater, the increase in this length was slight at 49.02 cm, whereas in the absence of salt, the plants recorded the longest root axis at 53.20 cm. In general, increasing the salt content of the substrate causes the main root axis to shorten (r= -0.474).

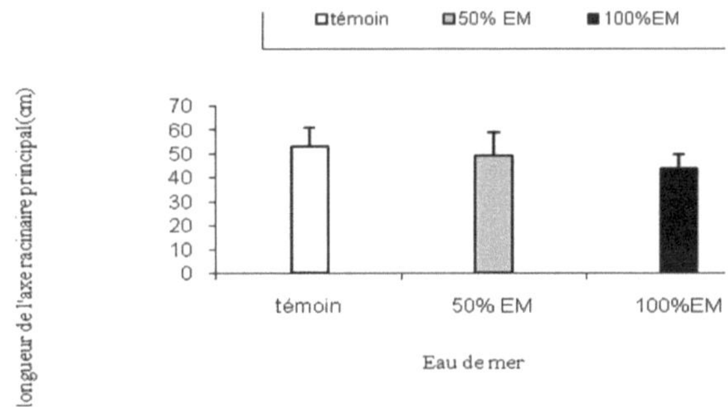

Fig.19 - The length of the main axis after 4 weeks of stress at Atriplexhalimus L. plants aged 120 days in **seawater.**

D-Anatomical parameters 1-the leaf

C4 plants have a special leaf structure. Microscopic observation of the leaf blade of Atriplexhalimus L. reveals two types of chlorophyll cells: those of the mesophyll, often characterised by the presence of calcium oxalate crystals, and a perivascular sheath surrounding the conducting bundles. In the centre of the latter are large cells without chlorophyll, which make up the aquifer parenchyma. The anatomy concerned in this study relates on the one hand to the Diameter of elements conductors represented by the vessel metaxyelic and the dimensions of chlorophyll cells,on the other.

*** Mesophyll compaction**

- to NaCl+CaCl₂

The expected results showed no notable significance for the measurements made (p>0.1). This proves that fluctuations in substrate salt concentration have only a slight influence on all the characteristics measured, except for the width of the sheath cell. Although it is the most affected by the imposed stress compared with the other measurements, its variations are only slightly dependent on the salt regime (p>0.05).

Table 9 - Statistical test of significance at P =1% for anatomical parameters of leaves of Atriplexhalimus L. plants stressed at **NaCl+CaCl 2 .**

Variables	Test F	probability	Mean and standard deviation		
			witness	400meq	600meq
Comp.meso µm	0.734	0.519	319±52.45	291±38.95	270±56.96
Diamxylm µm	0.419	0.675	16.11±2.07	14.39±1.83	15.05±3.25
Long cell.gain µm	0.508	0.625	43.88±2.16	37.92±12.68	37.93±6.61
Largcell.gain µm	3.930	0.081	30.29±2.05	22.52±4.66	25.11±3.14
Ep.mesofvt µm	2.767	0.141	55.99±10.10	62.25±11.20	41.14±12.44
Mesofd thickness µm	0.621	0.569	72.74±15.99	62.78±10.25	69.12±2.69

Meso compaction = mesophyll compaction

Diam xyl m =diameter of the xylemic vessel

Long cell. gain =length of sheath cell Lar gcell. gain =width of sheath cell
Ep.mesof vt = thickness of the mesophyll (ventral side) Ep.meso fds= thickness of the mesophyll (dorsal side).

The mean data (Table 9) show that the importance of chlorophyll parenchyma increases inversely with substrate salinity levels (r= - 0.442).

To this end, we note that in terms of the treatment carried out without salts, plants have the widest mesophyll with99.03 µm. In the 400meq treatment, its thickness is estimated at 90.66 µm, while in the 600meq treatment it is only 84 µm.

-to seawater

The same observations were made in this experiment. Analysis of the results shows that in all situations, the nature of the salt does not cause any striking changes in the leaf tissue concerned.
Examination of the results (Table 10) shows that changes in salinity levels are accompanied by slight oscillations in the magnitude of this variable ($p>0.1$).

Table 10 - Statistical test of significance at P =1% for anatomical parameters of leaves of Atriplexhalimus L. plants stressed with seawater.

variables	Test F	probability	Mean and standard deviation		
			witness	50% EM	100% EM
Compméso µm	1.697	0.260	223±9.83	316±85.33	241±74.89
Diamxylm µm	0.636	0.862	15.77±0.69	15.33±0.88	17.00±3.051
Long cellgain µm	0.591	0.583	29.51±5.38	31.84±6.63	33.91±0.89
Largcellgain µm	3.309	0.082	19.67±1.61	20.84±1.46	24.59±3.23
epomeofvt µm	0.482	0.639	73.34±16.63	63.68±16.50	59.03±21.11
epomeso fds µm	1.462	0.304	72.75±4.68	62.65±12.09	53.07±20.68

Analysis of the data shows that mesophyll appears to be very low in the control treatment, at 223 µm. In the treatment with 50% seawater, the expression of this parameter is more pronounced. At this level, the plants recorded a maximum value estimated at 316 µm, whereas at the treatment with 50% seawater, the plants recorded a maximum value estimated at 316 µm, whereas at the treatment with 50% seawater, the plants recorded a maximum value estimated at 316 µm.In the 100% seawater treatment, this measurement was evaluated at 241 µm, a value rather higher than that recorded in the control plants. This shows that the increase in salt content is accompanied by the formation of a more or less large mesophyll ($r=0.111$).

*** The diameter of the metaxylemic vessel :**

-to NaCl+$CaCl_2$

Looking at the average results (Table 9), the control plants recorded a diameter of 16.11 µm, which showed a slight decrease in the plants treated with 400meq 14.39µm and then stabilised at a level very close to that recorded in the control (15.05 µm). In this case, it should be noted that the increase in salinity only slightly reduces the diameter of the metaxylemic vessel.

-to seawater

This characteristic appears to show similar trends when compared with the previous measurement. The study of the results suggests that the diameter of the metaxylemic vessel shows small variations in magnitude linked to the concentration of salts in the irrigation water ($p>0.1$). A decrease in the salt content of the substrate is accompanied by a decrease in the diameter of the metaxylemic vessel ($r=0.296$). The average results show that the diameter shows the same value for the treatment carried out without salts and that carried out with 50% seawater with 15.77 µm and 15.33 µm, on the other hand the plants grown in heavily salted water, record a diameter estimated at 17.00 .

*** Sheath Cell Length**

- to NaCl+CaCl 2

The results obtained (Table 9) show that the length of the sheath cells reached a maximum value in the absence of salts, at 43.88 µm, whereas in the salinity treatment, the lengths were equivalent and appeared to be shorter than in the previous treatment (37.93) cm. Furthermore, the greater the salinity concentration, the shorter the length ($r=-0.330$).

-to seawater

During this measurement, the results reappear with the same tendency, showing that this parameter shows slight vacillations following variations in the salt regime ($p>0.1$).

A study of the average results shows that the control plants show a higher level of reduction with 29.51µm; irrigation with seawater at two different concentrations leads to an increase in this length. Thus the plants recorded an estimated value of 31.84µm when they were watered with slightly brackish water and 33.91µm when the water was heavily salted. Moreover, the higher the salt concentration, the greater the length ($r=0.406$).

The width

-to NaCl+CaCl$_2$

The measurements available to us (Table 9) show that cell enlargement reached an estimated maximum value of 30.29 µm in the treatment carried out without salts. Irrigation with salt water, on the other hand, results in a reduction in cell size, with 22.52µm and 25.11µm recorded in the 400 and 600 meq treatments respectively.

-to seawater

The statistical study of the results of the evolution of this characteristic (table10) shows that variations in salinity levels have only a slight influence on the fluctuations observed. However, the latter remain relatively significant in the expression of this width, when compared with those of previous measurements ($p>0.05$). The average results show that the regression of this width is reported to be of the same intensity in the treatment conducted without salts and that conducted with 50% seawater (19.67µm and20.84µm). However, the plants treated with 100% seawater clearly stand out from the other treatments, with a much greater width of 24.59µm. The increase in the salt content of the substrate was accompanied by a significant increase in the width of the sheath cell ($r=0.719$*).

*** Importance of the mesophyll Ventral side**

- to NaCl+CaCl 2

The results in Table 9 indicate that the variations in the salinity levels imposed allowed only very slight fluctuations in this characteristic. For theThe plants recorded 55.99 µm, a value that increased slightly in the 400 meq treatment (62.25 µm), but at a concentration of 600 meq this measurement decreased considerably compared with that of the first two treatments (41.14µm). The reduction in irrigation water salinity is accompanied by an increase in this parameter ($r=-0.474$).

-seawater

The results show identical observations, indicating that the salinity level seems to have a very small effect on this parameter on both sides ($p>0.1$).The average data (Table 10) show that the highest value was recorded in the control plants, with 73.34 µm. This value decreased slightly under saline conditions: 63.68µm and 59.03µm were recorded for the treatments carried out with 50 and 100% seawater, respectively.

Dorsal side

-to NaCl+CaCl$_2$

A reading of the average results (table 9) shows that this characteristic fluctuates slightly as a result of variations in the salt regimes adopted. Plants grown without salts recorded a maximum value of 72.74µm. Watering with saline solution reduces this value. Thus, 62.78µm was recorded in plants grown at400 meq and 69.12 µm in those grown at600 meq. Although this value is still higher than that of the previous treatment, it is lower than that of the control.A weak negative relationship was noted between the increase in salinity and the size of the mesophyll (r=-0.168).

-seawater

The results obtained show that the seedlings grown without salts have a very large mesophyll (72.75 µm) on their dorsal surface. This size is reduced in the presence of salts, where the differences between the control treatment and the other two treatments are sufficiently large, with values of around 62.65µm for the 50% seawater treatment and 53.07µm for the 100% seawater treatment. The reaction of the plants follows the same trends for the expression of this parameter on both sides, so its development intensifies with the degree of salinity.

2 - The root

In Chenopodiaceae, the internal structure of the root is slightly different from that found in other species. It is characterised by the presence of an abnormally functioning cambium, consisting of a large meristematic zone producing conductive tissue on its inner surface.
The characterisation of anatomical mechanisms was based on a structural study of the cortical parenchyma and the central cylinder. Observations focused in particular on cell dimensions and the number of cribrovascular bundles.

Cortical parenchyma thickness

-to NaCl+$CaCl_2$

Statistical analysis of the average results (table 11) shows that the thickness of the control plants was 22.05µm, while this value did not vary significantly in the plants treated with 400 meq, where a value of 21.55 µm was recorded. The subsequent increase was slight, with a value of 22.22 µm.
These average results show that the formation of cortical parenchyma does not seem to be affected by salinity. This is confirmed by the absence of any significant relationship (r^2 =0.019).

-seawater

Analysis of variance shows that variations in salinity levels have a greater or lesser influence on the development of this parameter ($p>0.05$).

The average results (Table 12) show that the cortical parenchyma is thicker in the control treatment (22.05 μm). When irrigated with salt water, all the plants reacted by reducing their cortical parenchyma. In the 50% seawater treatment, the thickness reached 14.08 μm and 12.78 μm in the 100% seawater treatment. These results show that increasing the degree of salinity causes a reduction in cortical parenchyma ($r=0.631$).

Table 11- Statistical significance test at P =1% for anatomical parameters of roots of Atriplexhalimus L. plants stressed with **NaCl+CaCl 2**

variables	Test F	probability	Mean and standard deviation		
			witness	400 meq	600 meq
E.p.c	1.982	0.018	71.01±11.93	69.00±11.93	71.00±7.76
L.c.p	0.705	0.531	30.68±5.25	32.62±12.70	25.24±3.18
L.c.p.	0.270	0.772	22.00±5.84	23.56±9.20	19.67±2.94
D.v.mx	5.246	0.048	15.53±4.47	25.24±5.71	15.27±1.61
Nbref.c.v	3.735	0.088	8.00±2.64	10.00±1.73	5.00 ±1.15
E.rhiz.	1.371	0.323	48.30±4.91	46.5 ±16.24	60.46±10.14

E .p.c = thickness of cortical parenchyma

L.c.p = length of the parenchymal cell

L.c.p = l argeur of the parenchymal cell

D. v. mx = diameter of the metaxyemic vessel

Nbref. c. v = number of criblo vascula ire bundles

E. rhiz = rhizoderm thickness

*** Number of beams**

-to NaCl+CaCl₂

An analysis of the results shows that salinity had no effect ($p>0.05$) on the variations observed. The increase in salinity did not significantly affect the number of clusters ($r= -0.401$).

The average results show that the 400 meq treatment recorded a higher number of clusters, estimated at 10, while the number of clusters decreased in the control and 600 meq plants, where 8 and 5 clusters were observed respectively.

-to seawater

An analysis of the results obtained from this measurement shows that variations in this measurement are closely linked to the salinity level ($p<0.05$).Analysis of the average results from this estimate reveals that the degree of salinity has considerably reduced the number of clusters formed.It is therefore clear that the imposed salinity causes a change in root structure ($r=- 0.750$).*In the control plants, 8 clusters were recorded, whereas in the 50% and 100% treatments, only 2 clusters were observed.

*** Diameter of the metaxylemic vessel**

-to $NaCl+CaCl_2$

Analysis of the results obtained (table 11) shows that variations in salinity levels cause significant oscillations in this parameter, which means that any reduction in the salt content of the substrate is accompanied by a reduction in diameter ($r=-0.019$).The average results show the same diameter for the control plants and those conducted at 600 meq (15.53, 15.27 µm), while the treatment conducted at 400 meq recorded a maximum value estimated at 25.24 µm.

-seawater

In contrast to the previous case, seawater spraying at different doses caused only slight fluctuations in this parameter ($p>0.1$). The increase in salinity was accompanied by a reduction in the diameter of the metaxylemic vessel ($r=-0.338$).A study of the average results shows that, across the treatment carried out without salts, the plants recorded the largest diameter.At 15.53 µm, this diameter decreased with the saline treatments, reaching 11.39 µm in plants treated with 100% seawater and 8.93 µm in those treated with 50% seawater.

*** cell dimensions in cortical parenchyma**

A study of the results obtained when estimating the dimensions of the cortical parenchyma cell shows that the length and width of the cell are only slightly influenced in their development by the saline regimes applied ($p>0.1$). Thus, the influence detected was part of a negative trend ($r=-0.310$).

Table 12: Statistical test of significance at P = 1% for anatomical parameters of Atriplexhalimus L. plants exposed to **seawater**.

variables	Test F	probability	Mean and standard deviation		
			witness	50% EM	100% EM
E.p.c	2.991	0.126	71.01±11.93	30.23±24.02	41.14±24.93
L.c.p	6.304	0.034	30.68±5.25	16.05±7.02	20.84 ±7.02
L.c.p.	2.858	0.134	22.00±5.84	9.58±7.81	16.70±5.22
D.v.mx	1.26	0.360	15.53±4.47	8.93 ±6.72	11.39±3.76
Nbref.c.v	12.03	0.008	8.00±2.64	1.00 ±0.57	2.00 ±1.15
E.rhiz.	3.74	0.088	48.30±4.91	26.47±4.98	36.42±15.42

The length

-to $NaCl+CaCl_2$

The expression of the parenchymal cell length was more pronounced in the 400 meq treatment, with a value of 32.62 µm. A slight decrease in this value was observed in the control treatment, with 30.68 µm ;However, in the 600 meq treatment, the plants recorded the lowest value at 25.24 µm.

-seawater

The results show that this anatomical characteristic is strongly linked to the salinity levels imposed ($p<0.05$).A study of the average results shows that the longest cell, measuring 30.68 µm, was obtained in the control plants. This length was reduced in the saline regimes, reaching 16.05 µm in the 50% seawater treatment and 20.84 µm in the undiluted seawater treatment.

The width

-to $NaCl+CaCl_2$

Estimation of this width shows that variations in width effectively represent the same trend as that recorded when length was measured. As a result, the plants treated with 400meq recorded a maximum value of 23.56 µm. In the treatment carried out without salt, this length was reduced and recorded a value of 22.00 µm, whereas in the treatment carried out with 600meq, it only reached 19.67 µm.

-seawater

The results show that the control plants have a wider cell. Under saline conditions, the width was greatly reduced, with 9.58 µm recorded in plants fed 50% seawater and 16.70 µm in those fed 100% seawater.

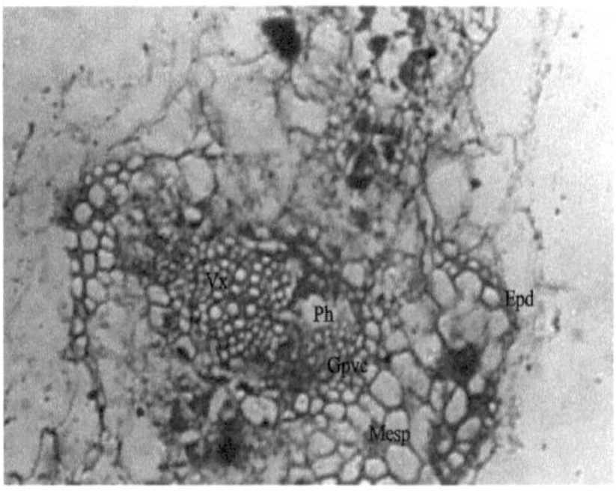

Fig.20- Anatomical section illustrating the structure of the leaf (Control). Epidermis: Epidermis. Mesp: Mesp. Gpvc: Perivascular sheath.Ph: Phloem. Vx: Xylemic vessel.

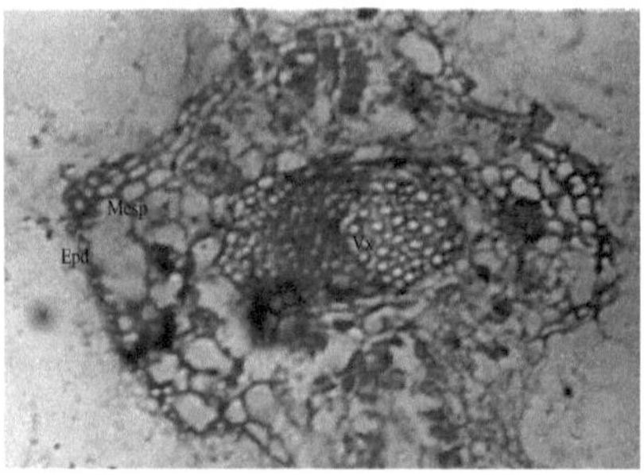

Fig.21- Anatomical section illustrating the structure of the leaf (400 meq).

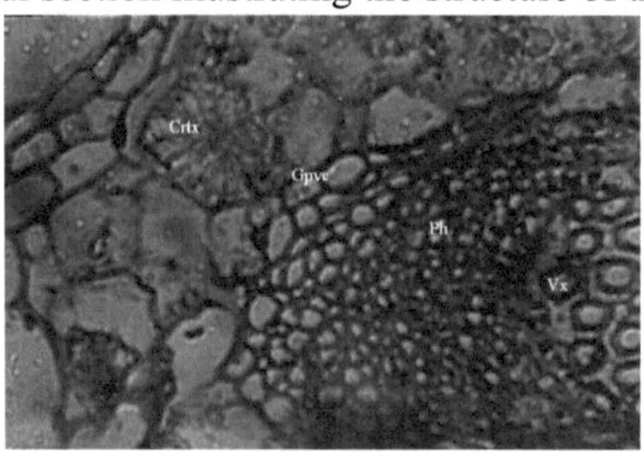

Fig.22- Anatomical section illustrating the structure of the leaf (600 meq). Crtx: Calcium oxalate crystals.

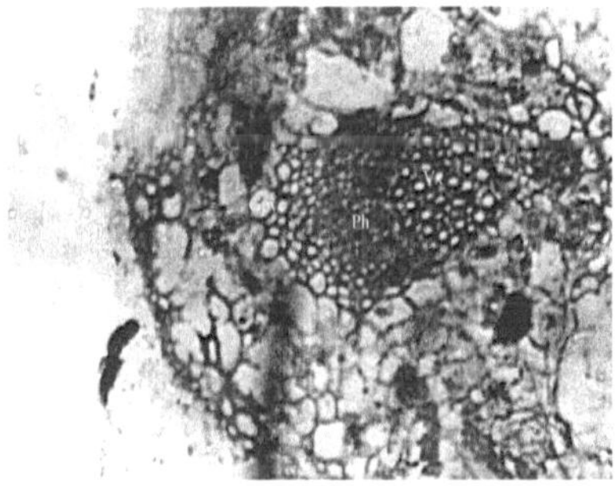

Fig.23- Anatomical section illustrating the structure of the leaf (50% EM).

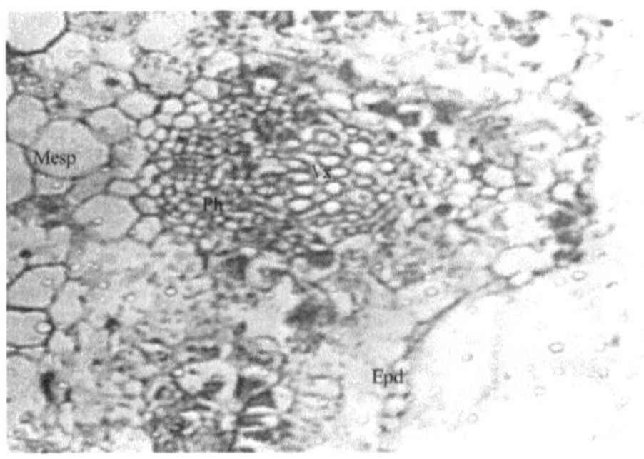

Fig.24- Anatomical section illustrating the structure of the leaf (100% EM)

DISCUSSIONS AND GENERAL CONCLUSION

From the results we have just commented on, the following key points emerge:

- Relative water content is considered to be an excellent indicator of the plant's water status. It is a parameter that can be greatly influenced by any variation in the plants' absorption potential.

However, all the conditions that contribute to variations in substrate potential, such as salt and water stress, are largely reflected in the estimation of this water characteristic.Such observations have been widely reported in numerous studies, which show that increasing salinity in the growing medium considerably hampers water uptake by the plants in place. In this context, Khan et al (2000) show that an increase in salinity results in a drop in tissue water content, with the onset of osmotic adjustment mechanisms.The results of this experiment (Table 13 in the appendix) show that the installation of salt stress depends closely on the salt stress regime used (NaCl+CaCl2 and diluted or undiluted seawater). The data obtained indicate that the relationships that emerged were very different.The application of increasing concentrations of seawater has a negative and highly significant influence on the depression of tissue water content during the first phase of its estimation (r=-0.761*). This influence is minimised at a later stage of the experiment. However, the negative influence remained, but was estimated at a lower intensity (r= - 0.396). These results seem to indicate that under salinity express adaptation mechanisms that maintain a constant optimal hydration. The state of stress caused by the application of the NaCl+CaCl combination2 shows slightly different trends. The concentrations used had no significant effect on the evolution of the water content of the tissue. This influence is very slight and a positive trend (r=0.244 and r=0.003).These relationships indicate that, under NaCl+CaCl salt stress2 , water availability is higher than in the seawater irrigated system. According to some studies, the addition of organic matter to the growing medium reduces the stressful effect of salts on the potential for water uptake by existing plants (Hannachi et al., 2004). The role of organic matter is linked to its capacity to adsorb cations released by the solubilisation of salts. Thus, in our situation and still under NaCl+CaCl2, , the cations emanate solely from the passage of the salts used into solution.Unlike the batch irrigated with seawater, these cations are more numerous and diversified. Comparing the two situations, the water retention forces, mainly osmotic, are greater.With regard to the relationship between relative water content and wax and transpiration rates, the behaviour of plants in relation to salinity varies according to the salt regime. The nature of the salts used appears to be the source of divergent wax deposition behaviour. A study of these relationships (tables 13 and 14 in the appendix) clearly indicates that the loss of turgidity in the seawater-treated system is accompanied by a fairly significant deposition of epicuticular waxes (r=-0.655). However, noNo significant reaction was observed in plants treated with NaCl+CaCl2 on wax synthesis and deposition (r=0.192).While the accumulation of waxes in the plants of the species studied may be a consequence of the perception of the restrictive effects of salt stress, irrigation with seawater has harmful effects on the plants compared with irrigation with NaCl+CaCl .2From these results, it should be noted that the applied salinity leads to a very pronounced reduction in the levels of water emission by the excised leaf (RWL) that we simulate as representative of global transpiration under in situ conditions. These negative and significant relationships therefore show little fluctuation under the effect of the two salinity regimes.For plants treated with seawater, the interaction between salinity and water emission was higher (r= -0.801**) than that recorded under NaCl+CaCl2 (r=-0.707*).In addition, a study of the relationships established between transpiration intensity, estimated by water loss through the excised leaf, and wax deposition reveals that these two parameters are significantly linked, with a negative trend.This explains why the wax deposition observed under salinity conditions is the cause of a pronounced limitation in transpiration, essentially cuticular transpiration observed under both types of stress. This relationship was more significant with the application of NaCl+CaCl2 (r=-895**) than with seawater irrigation (r= - 0.719*). These results are in agreement with those of Cotoole et al (1983); Austin (1987); Acevedo (1987) and Dib (1992) cited by Adda in 1996, showing that a reduction in leaf surface accompanied by the deposition of epicutucular waxes limits transpiration. These transformations allow the leaf to maintain hydration for optimal physiological function.A study of the parameters relating to the control of water loss shows that transpiration measured on the excised leaf decreases with salinity, regardless of the type of salinity (r= -0.801** for seawater and -0.707 * for NaCl +CaCl 2). In this context, stomatal resistance also increases (r=0.289 for seawater and 0.327 for NaCl +CaCl2).This reaction is obvious, as the depletion of water from the tissues as a result of variations in the osmotic potential of the substrate due to the high salt content leads to the closure of the stomata, inhibiting stomatal transpiration.This response is often managed by the availability of water in the substrate. Accumulated salts, not absorbed by the plant, act by increasing the osmotic pressure of the substrate, which prevents absorption and reduces the hydraulic conductivity of the roots, causing a reduction in the flow of water from the roots to the aerial parts (Leary, 1969; Prisco, 1980 cited by Azevedo, 2004).This behaviour seems to reflect a similarity in the action of salinity with those of drought.The relationship between salinity levels and above-ground and below-ground biomass (Tables 15 and 16 in the Appendix) shows that morphological changes in the root system depend closely on the degree of salinity imposed on the growing substrate.The biomass deficit recorded in plants subjected to salinity caused by two types of salt did not affect the two parts of the plant in the same way.In fact, root growth was less affected by salts than above-ground growth. These results are confirmed by the work of Da Silva et al. in 2003, which showed that salinity seems to induce a preferential allocation of biomass to the roots, the greater the intensity of the stress. This is shown by the fact that the rate of reduction in above-ground biomass of plants grown in the presence of salts is much higher than that of control plants (r= -0.686* for seawater and -0.762* for NaCl+CaCl).2The allocation of biomass to the roots also resulted in higher PR/PA ratios in the most stressed plants (r=-0.944** for seawater and r=-0.241 for NaCl+CaCl).2The development of the root at the expense of the aerial part is considered by several authors to be a criterion of resistance to salinity. It allows water to be extracted from salty environments and retained, thereby maintaining a higher relative water content (r=0.770*).With regard to the length of the main root axis, plants do not react in the same way to the stress imposed. The two types of salinity differ in the expression of this characteristic. Irrigation with seawater leads to a shortening of the main axis (r= - 0.698*), whereas irrigation with the Na Cl+CaCl solution2 , favours root elongation (r=0.259).Such a relationship could be explained by the inhibition of growth due to the nature of the ions found in seawater, which diminishconsiderably the water potential of the soil solution by imposing real water stress conditions.This result is in line with that of (Azevedo et al., 2004) who prove that salt stress considerably affects the growth of the root system. Such morphology would not limit variations in the water content of leaves (r=0.197 for seawater, r=-0.153 for NaCl+CaCl).2The relationships established between relative water content and sugar content show that the reduction in sugar content in the leaves favours the maintenance of an optimal state of turgidity (r=-0.382). As water retention forces in the substrate increase during stress, osmotic adjustment may occur, but to varying degrees.This result has been confirmed by numerous studies which have shown that salt stress is accompanied by an accumulation of osmoticum (Clandivan et al., 2003; Da Silva et al., 2003), indicating osmoregulation, which enables leaf cells to maintain a level of turgidity compatible with metabolic activities. Furthermore, the relationship between the level of sugars measured in a second phase and the relative water content was significant (-0.411).This relationship could be explained by the hypothesis that the plants react to the severity of the stress by progressively increasing the sugar content in the leaf tissue to overcome the progressive increase in the salt content of the environment. The evolution of sugar content as a function of salinity caused by the application of a NaCl+CaCl solution2 follows a similar trend to that caused by the application of seawater, but at a lower level. This could be explained by the the deleterious effect of sodium is reduced by increasing Ca concentrations.+2According to Hopkin, 2003 it appears that Na+ displaces Ca+2 bound to root membranes, an effect that can be avoided by supplying excess Ca+2 resulting in a decrease in Na uptake .+The relationship between osmotic potential and relative water content led to the expression of a drop in the osmotic potential of the leaves, particularly during the second measurement (r= -0.588 and r=- 0.918).*This lowering of the leaf osmotic potential would allow a high relative water content to be maintained (r= -0.544, r=- 0.191).this relationship, although weak, shows that the relative water content would not be attributed to the accumulation of osmoticums.With regard to the anatomical study of the leaf, the observations made (Table 17 and 18 in the appendix) indicate a reduction in the thickness of the mesophyll, favouring the maintenance of an optimal state of turgidity expressed by its RWC (r= -0.429 for seawater and r=-0.224 for NaCl+CaCl2). This reduction would allow less water to be lost, thereby maintaining a high water potential.For its part, the characteristic linked to the diameter of the metaxyemic vessel would undergo the same reactions. Vessels with a smaller diameter would be responsible for a higher water content (r=-0.326 for seawater and r=-0.466 for NaCl+CaCl).2Such a transformation would facilitate the flow of sap, in conjunction with a lower water content in the substrate due to the accumulation of ions.At the perivascular sheath cell level, observations showed that under seawater irrigation, the increase in relative water content led to an increase in cell size in terms of length (r=0.699*) and width (r=0.474). This increase in sheath cell size probably contributes to the regulation of photosynthetic intensity. This increase in sheath cell size probably contributes to the regulation of photosynthetic intensity. However, the effect on these anatomical parameters was clearly attenuated when the plants were sprayed with the Na Cl+CaCl solution2 (r=-0.259 length, r=-0.298 width).In addition, the effect of salt stress, either seawater or NaCl +CaCl2 caused a reduction in chlorophyll parenchyma on both ventral (r=-0.474) and dorsal (r=-0.149) sides for Na Cl+CaCl2 and under seawater for both sides respectively (r=-0.365 fv and r=-0.572 fd). This decrease can be explained by the reduction in mesophyll cell size. The effect of salinity is a reduction in the availability of water to the plant, due to an increase in the osmotic pressure of the substrate. This reduction leads to a loss of turgidity.In this case, the structural transformations probably consist of a reduction in the protoplasmic content sensitive to dehydration. Our results show that the relative water content is maintained in plants with a smaller mesophyll (r=-0.305 for the ventral surface, -0.381 for the dorsal surface under NaCl+CaCl2 and r=-0.0.024 for the ventral surface and r=0.242 for the dorsal surface under seawater).In the root system, watering with seawater causes a reduction in the thickness of the cortical parenchyma (r= -0.500). This reduction is conditioned by the length (r=0.825**) and width (r=0.799**) of the cells making it up, which in turn are reduced by the imposed salinity (r= -0.543 length, r=-0.297 width).The reduction in cell size and, consequently, the thickness of the cortical parenchyma would be linked essentially to the lowering of turgidity, which would imply the limitation of cell expansion. This result confirms that of Beatriz et al. in 2001; Yang Hui Peng et al., 2004, showing that saline stress penalises cell expansion. However, irrigation with different concentrations of NaCl+Ca Cl2 had no significant effect on the development of this structural characteristic (table 19 in appendix).This can be explained by the absence of any significant relationship between the two variables (r=0.019 Epc, r= -0.310 Lcp, r=-0.171 Lacp).The absence of the saline effect could also be explained by the concentration of Ca++ in the tissues increasing sharply when this cation becomes higher in the medium. Indeed, Levitt, 1980 and later Alem et al., 2005, report that under saline stress conditions, the addition of calcium to the substrate improves plant tolerance by promoting growth. Furthermore, according to Da Silva et al. 2003, this cation could limit the entry of Na+ and maintain selectivity between K+ and Na+ at root level.It should also be noted that seawater salinity stress has a highly significant negative effect (table 20 in appendix) on the number of libero-ligneous bundles (r=-0.701*).In this way, high concentrations of substrate salts are accompanied by a reduction in the number of beams.Watering with saline solution containing Na Cl+Ca Cl 2, the relationship established between the two factors, although weak, follows the same trends (r=-0.401).The diameter of the metaxylemic vessel appears to be reduced by the salinity imposed by seawater (r= -0.338). However, the expression of this parameter is unaffected by the salinity regime caused solely by Na Cl+Ca Cl 2 (r=-0.018).This modification intensifies sap conduction and adjusts absorption capacity in relation to the humidity of the substrate conditioned by the high temperatures.salt concentrations created by the accumulation of sulphates, hydrocarbonscarbonates or a mixture of several water-soluble salts of Wed. Another characteristic was retained in this study, relating to the thickness of the outermost layer of the root represented by the rhizodermis. The relationships established show that the stress imposed varies to a greater or lesser extent in terms of the expression of this modification. Thus, the increase in saline concentrations caused by seawater irrigation led to a thinning of the rhizodermal layer (r=-0.409). On the other hand, the addition of the NaCl+CaCl solution2 led to a thickening of this layer (r=0.441).The thickening of the rhizodermis is thought to be a reaction aimed at reinforcing the barrier against the dissipation of water from the roots into the soil and maintaining turgidity.The results show a diversity of behaviour linked to the nature of the stress and the salt concentration.In view of the data we have just concluded, in order to select salinity-tolerant and/or salinity-resistant atriplex species, it is possible to :

- To look for interesting traits that provide additional elements likely to explain the mechanisms of tolerance to abiotic stresses, particularly saline stress,
- Research into molecular markers that can be used for stress response,
- Helping to identify the metabolic pathways involved in the stress response.

REFERENCES

ADDA A., 1996 - Etude des paramètres morpho physiologiques et anatomiques du blé dur, dans une zone semi -aride. Thèse de Magister. INA El Harrach Alger, 120 p.

AJAB M., 2002 - Recherche de traits morphologiques physiologiques et bichimiques d'adaptation au déficit hydrique chez différents génotypes de blé dur Triticumdurum Desf. Thèse de Magister. Univ Annaba, 84 p.

ALEM C., AMRI A., 2005 - Importance of cell membrane stability in salinity tolerance in barley. Biology and Biotechnology.Vol.4, (1), p. 20 -31

ALARCHON JJ., SANCHEZ-BLANCO MJ., BOLARIN MC,

HIGAZY SHEHATA M., and ALLAM A., 1993 - Free proline relation to salinity tolerance of three sugarbeet varieties.Egypt. J. Agric.Res. 73, (1), p 175 -189.

AL HAKIMI A., 1991 - Recherche de caractères morphologiques d'adaptation au déficit hydrique chez les espèces tétraploïdes de blé. D E S. Univ. Monpelier II.110 p.

ALI DIB T., MONNEVEUX P and ARANS, JL., 1992.
Adaptation to drought and the notion of idiotype in durum wheat II. Physiological characteristics of adaptation. Agronomie, 12, p 381 -393.

AMRAR S., 1993 - L'amélioration de la tolérance à la salinité chez la tomate Licopersicumesculentum. Master of Science, C.I.H.E.A.M Zarragoza. Spain, p7 -23.

AOUISSAT M., 1992 - Utilisation des arbustes et des ressources fouragères dans la production ovine extensive. Magistèr thesis, C.I.H.E.A.M Zarragoza.Spain110p.

AUBERT G., 1975 - Sodic soils in North Africa. Annex. Agro, I.N.R.A. Alger, p 4 -5

AUBERT G et BOULAINE J.,1976 - La pédologie. Collection "que sais je". Ed. I.P.V de France. p38 -42.

AUBERT G et BOULAINE J., 1980 - Pédologie. Ed. I.P.V de France.p25 -35.

AZEVEDO A., JOSE T., JOAQUIM EF., CLAUDIVAN F.,
VIEIRA SILVA J., PAULO H, and ENEAS, 2004 - Effects of salt stress on plant growth, stomatal response and solute accumulation of different maize genotypes. Braz. J. Plant Physiol. vol.16 no.1.

BATANOUNY H., 1993 - Ecophysiology of halophytes and their traditional use in the Arab world. Advanced course on halophyte utilization. Agriculture, p12 - 23, Agadir. Morocco.

BENKHETOU A., 2003 - Contribution à l'étude de la mise en culture des zones steppiques dans le cadre de l'accession à la propriété foncière agricole et son impact sur l'écosystème (cas de rechaiga-Tiaret). Thèse de magister 98p. University of Tiaret.

BERNSTEIN N., LÄUCHLI A and SILK WK, 1993 -
Kinematics and dynamics of sorghum (Sorghumbicolour. L) leaf development at various Na/Ca salinities. I. Elongation and growth. Plant Physiol, 103: p 1107 -1114

BIDINGER FR., WITCOMBE JR. 1989 - Evolution of specific hydration tolerance traits for improvement of drought resistance in F.W.G, bakerled drought resistance in cereals, CAB international, willing VK, PP151 -164.

BINET P et BRUNEL JP, 1967 - Physiologie végétale (I). Ed DOIN Derven et Cie.

BOURSIER P. et LAUCHLI A.,1989 - Mechanisms of chloride partitioning in the leaves of salt stressed sorghum bicolour L.Plant physiology.77p.537 -544.

BRESSAN and PAUL M HASEGAWA, 1993 - NaCl
Regulation od Plasma Membrane H+-ATPase Gene Expression in a Glycophyte and a Halophyte. Plant Physiol. 103: p 713 -718.

BRUGNOLI E and BJÖRKMAN O., 1992 - Growth of cotton under continuous salinity stress: Influence on allocation pattern, stomatal and non-stomatal components of photosynthesis and dissipation of excess light energy. Planta, 187 : p335 -347

CAMEFORT H., 1984 - Morphology of vascular plants.
Ed:Doin, Paris, 432P.

CHEVERRY C., 1995 - Plant behaviour in saline environment. Action eau N°4 séance spécialisée du 22 mars 1995.Ed.Acad agr, France,49p.

CHRETIEN D., 1992 - La résistance au sel chez le jojoba (Simmondsia chinensis LS.) :croissance et modifications du contenu lipoprotéique de cals cultivés en présen ce d'une teneur élevée en NaCl. Thèse de Doct.Univ.,Paris,144 p.

CHRISTOPHE M et MICHEL T., 2000 - Statut hydrominéral des plantes. Ed TEC et DOC. p 92 -118.

CLARCKE J and Mc CAIG T., 1982 - Excised leaf water retention capability as an indicator of drought resistance of Triticumgenotypes. Canadian j.plants.sci. (62) :p :571 -578.

CLEMENT M., FRANÇOISE P., 1998 - Analyse physique sols.
Méthodes choisies. Ed Lavoisier :TEC et DOC

COLMER TD., FAN TW-M., HIGASHI RM. and LÄUCHLI A.,
1996 - Interactive effects of Ca2 + and NaCl salinity on the ionic relations and

proline accumulation in the primary root tip of Sorghum bicolour . L. Physiol. Plant. 97:421 -424.
DEBEZ A ., CHAIBI S ., BOUZID, 2001 - Effect of NaCl and of growth regulators on germination of Atriplexhalimus L. Agricultures. Volume 10, N°2, p 135 -138.
DERAISSAC M., 1992 - Mécanismes d'adaptation à la sécheresse et maintien de la productivité des plantes cultivées. Agro-Tropical, 64 (1) p23 -37.
DIGBY C., SHORT and TIMOTHY D. COLMER, 1999 - Salt Tolerance in the Halophyte Halosarciapergranulata sub sp.Pergranulata. Annals of Botany 83:p 207 -213.
DOMERGUEZ M and MANGENOT R., 1970 -Ecology microbial. Academic press NEW YORK, 395 p.
DUCHAUFOUR P., 1968 - L'évolution des sols, essai sur la dynamique des profils. Ed Masson et Cie. p93.
DUCHAUFOURP.,1983 -Pedology 1.Pedogenesis et classification. Ed Masson. Pp 468 -470.
DUCHAUFOUR P., 1988 - Abrégés de pédologie. Ed Masson. Pp 197 -198.
DUCHAUFOUR P., MAURICEB et BERNARDS S., 1979 - Pédologie 2 "Constituants etpropriétés des sols"; Ed. Masson; Paris.
DUPERATM.,1997 -The guide des trees and shrubs from France. Ed. Sélection du reader's Digest.255p.
EILERS R G., EILERS W D and LELYK G. 1995 - Santé des soils. Sécherèsse. Ed john libbeyeurotext, Canada, pp 23-33 .
EL AICH A., 1987 - Fodder trees and shrubs in range and farming systems in North Africa. Malaysian Agricultural Research and Development Institute (MARDI),P 92 -96.
ELENA VV., ELENA GA., VINCENTR F., VLADIMIRI P., LAVIKIIRATSO., MAURICES KU and GERALDE E.,2001 - Salsolaarbusculiformis,a C3 -C4Intermediate in Salsoleae (Chenopodiaceae). Annals of Botany 88:337 -348.
EPSTEIN E., 1998 - How calcium enhances plant salt tolerance. Science 40:1906 -1907.
F A O., 1989 - Bilan préliminaire de 30 ans de pastoralisme, technique de développement pastoral.
FRANCLET A et LE HOUEROU H., 1971 - Les Atriplex en Tunisie et en Afrique du Nord. Rome: Food and Agriculture Organization of the United Nations (FAO), 271 p.
GAN Y., LAFOND GP and MAY WE., 2000 - Grain yield and water use:

Relative performance of winter V.S. Spring cereals in cost central Saskatchewan .Canadianjournal of plant science (533 - 541).
GASMI A., 1989 - Amélioration des propriétés physiques des sols salés par l'action combinée de la paille de blé et de phosphogypse. Mémoire d'ing agro, Univ de Tiaret.
GATE P., 1995 - Ecophysiologie du blé, Ed: Lavoisier, Paris, Techniques et Documentations, 429 p.
GAYRAL P., VINDT J., 1961 - Anatomie des végétaux vasculaires. Ed Doin et Cie. Paris p 114 -115.
GERARD R., 1993 - Métabolisme des végétaux, physiologie et biochimie. Ed. Presses Polytechniques et Universitaires Romandes. pp 27 -48.
GILBERT GA., GADUSH MV., WILSON C and MADORE MA .1998 - Aminoacid accumulation in sink and source tissues of ColeusblumeiBenth. duringsalinity stress. J. Exp. Bot. 49:107 -114.
GIRIJA C.,SMITH BN., SWAMY PM, 2002 -Interactive effects of sodium chloride and calcium chloride on the accumulation of proline and glycinebetaine in peanut (Arachishypogaea L.). Environ. Exp. Bot. 47:1 -10.
GUTTOUCHE R., 1998 -Contribution à l'étude des caractères morpho-physiologiques d'adaptation à la secheresse du blé dur(TriticumdurumDesf). Mémoire de Magister,INA El Harrach. Alger.120p.
HAMZA M., 1980 - Réponses des végétaux à la salinité.Physio- Vég.18, p 69-81.
HANNACHI C., DEBERGH P., EZZEDINE Z., MESSAÏ A et MEHOUACHI T. 2004 - Salt stress tuberization of potato (Solanumtuberosum L.) vitroplants. Biotechnol. Agron. Soc. 8 (1), p 9-13.
H C D S., 1996 - Note bibliographique sur quelques plantes fouragères et pastorale,p6 -7.
HELLER R.ROBERT E.,CLAUDEL.,1998 - Physiology vegetal 1 - Nutrition, Edit Duno, Paris, 322p.
HOPKIN WG, 2003 - Plant physiology. Ed De boeck. 514p.
HOUCHI R et COUDRET A.,1994 - Triticale selection tolérants au sel, Cahier Agriculture N°3,p.p 227 -230.
HSIAO TC. 1973 - Plant responses to water stress. Ann Rev Plant Physiol ; 24 : 519 -523.
HUBAC C.,1990 - Croissance et développement des végétaux. Impact de la salinité et l'aridité sur la croissance, le développement et l'amélioration des végétaux. ConférenceUniversitéd'OranEs- senia.
KATEMBE WJ,UNGAR IA,MITCHELL JP, 1998 - Effect of Salinity on germination and seedling growth of two Atriplex species

(Chenopodiaceae). Ann Bot; 82: p165 -75.
KHRAIST.,LECLERCY.,DONNELLYDJ,1998 -Relative salinity tolerance of potato cultivars assessed by in vitro
screening. Am. J. Potato Res. 75, p. 207-210.
KINET JM., BENREBIHA F., BOUZID S., LAILHACAR S., and
DUTUIT, 1998 - Atriplex network. Combining biotechnology and ecology for greater food security in Africa.Cahiers de l'Agriculture.vol(7) 6,p :505 -509.
KINRAIDE TB., 1999 - Interactions among Ca2+, Na+ and K+ in salinity toxicity: quantitative resolution of multiple toxic and ameliorative effects. J. Exp.Bot. 50:p1495- 1505.
KORICHI M., 1992 - Contribution à l'étude de l'ajustement osmotique de la réponse protéique du blé dur (Tritucumdurumdesf) au stress hydrique D.E.A. ENSA. Montpellier (France) P20 -35.
KRAMER PJ, BOYER JS, 1995 -Water Relations of Plants and Soils. AcademicPress, San Diego.
LACOSTE A and SALANON R.,1981 -Elements of biogeographies et d'écologie. Ed Nathan. pp119 -121.
LAFON JP,THARAUDL.PRAYERC.,LEOVG.,1985 -Biologie des plantes cultivées: Tome1 - Organisation physiologie de la nutrition. Ed L'ARPEPS ALGER.238p.
LADIGES PY, 1975 - Some aspects of tissue water relation in three populations of Eucaliptusviminalis Labill -New phytol, 75: p 53-62.
LAFON JP.,CATHERINE TH., PRAYERand GALLES L.,
1996 - Biologie des plantes cultivées, organisation physiologie de la nutrition.2 ème Ed. Tome 2.pp 135 -146.
LARSAM M., 1995 - Salinityproblems in the mediterranians area, N°2 séance spécialisée du 22 mars 1995.Ed.Academie de Paris,p.p 47 -59.
LAVAL-MARTIND.,MAZLIAKP.,1979 -TDandTP"
Plant physiology". Ed HERMANN collec. Méthod (Paris), 349p.
LAVAL-MARTIN D., MAZLIAK P.,1995 -Physiology
plant I. Nutrition and metabolism. Ed HERMANN.539 p.
LECLERC JEAN-CLAUDE, 1999 - Ecophysiologie végétale. Published by the University of Saint-Etienne, 35 rue du 11 novembre, 42023, 277 p.
LE GUYADER H., 1997 - Le développement des végétaux aspects théoriques et synthétique. Ed Masson pp 153 -156.
LEVIGNERON A., LOPEZ F., VANSUYT G., BERTHOMIEU P.,
FOURCROY P. et
CASSE-DELBART F., 1995 - Plants under salt stress.
Cahiers Agricultures , 4 : p263 -273.

LEVITT J., 1980 - Water stress- In Response of plants to environmental stresses. Vol II . 2ed Ed. by.T. Kozlowski. Academic Press: p 25 -29.
LOZE T et MATHIEU C., 1990 - Dictionnaire des sciences des sols 2ème Ed. Doc et Tec. Lavoisier, Paris 384p.
LUTTGE U., KLUGE M., BAUER G., 1997 - Botany 2 eme edition, Ed, Lavoisier Tec et Doc. Paris: 588 P.
MAAS EV., 1986 - Salt tolerance of plants. ApplAgricRes ; 1 : p12 -26.
MALLEK-MAALEJ E., BOULASNEM F., BEN SALEM M., 1998 - Effect of salinity on the germination of cereal seeds grown in Tunisia. Cahiers Agricultures; 2 : p153 -160.
M'BAREK B., CHAABANE R., SDIRI H., LAÏD M., 2001 - Effect of salt stress on germination, growth and grain production of some Maghrebian wheat varieties. Sécheresse ,12 :167 -174.
MELONI D., MARTA G., CARLOS A., 2004 - The effects of salt stress on growth, nitrate reduction and proline and glycinebetaine accumulation in Prosopisalba. Braz. J. Plant Physiol. vol.16 no.1 Londrina.
MOH'D KHAIR J., EL-SHATNAWI, 2002 - Dry matter accumulation and chemical content of saltbush (Atriplexhalimus) grown in Mediterranean desert shrub lands.New Zealand Journal of Agricultural Research, Vol. 45: p139-144.
MONNEVEUX PH et THIS D.,1997 - La génétique face aux problèmes de la tolérance des plantes cultivées à la sècheresse : espoirs et difficultés. Sécheresse 8(1) : p : 29 -37.
MOREL R., 1989 - Les sols cultivés. Ed Lavoisier: Tec DOC, p: 161-172.
MOUHOUCHE B et BOULASSEL A., 1999 - Contribution à une meilleure maîtrise des losses of irrigation water and salinisation of soils in zones arides. Recherches Agronomiques:(4) p15 -23.I.N.R.A.ALGER.
NEGRE R., 1961 - Petite flore des régions arides du Maroc Occidental. Tome 1, Ed, C.N.R.S.Paris.179 p.
NILSEN E.,DAVID MA. 1996 -The physiology of plants under stress. Abiotic factors Ed. JOHN WILEY and SONS. NEW YORK, p193 -210.
PHAM THI AT., BORREL-FLOOD P., VIERRA DA SILVA JB., JUSTIN AM andMAZLIAK P., 1985 - Effects of water stress on lipid metabolism in cotton leaves. AGRI.Bio.Chem.54 (3): P: 745 - 751.
POURRAT Y., DUTUIT P., 1994 - Étude précoce des effets morphologiques et physiologiques du rapport sodium calcium in vitro sur une population d' Atriplexhalimus . AUPELF-UREF ed. John LibbeyEurotext. Paris, p 293 -295.

RASCIO A., CEDOLA M., SORRENTINO GPASTORE D and WITTMER G., 1988 -Pressure volume curves and drought resistance in two wheat genotypes. Physiol Plant, 73 p 122 -127.
REDA TAZI M., BERRICHI A., HALOUI B., 2003 - Effect of the polyethylene glycol on germination and in vitro growth of the argan tree (Arganiaspinosa L. Skeels) in Beni Snassen (eastern Morocco). Sècheresse, vol 14 N°1 :3 -7 .
REKIKA D., 1997 - Identification des caractères physiologiques liés au rendement en condition de sécheresse chez le blé dur. Thesis: Doc. ENSA Montpellier 102p.
ROBERT D., CATERSSON AM., 1990 - Plant biology Volume 2. Characteristics and evolutionary strategy of plants. Vegetative organisation, p169 -170.
ROBINSON MF, VERY AA, SANDERS D and MANSFIELD TA, 1997 - How can stomata contribute to salt tolerance? Ann. Bot. 80: p 387 -393.
SCHUT P., 1996 - Manual acidity, salinity and solonetzic soil. Canola response tesoacidity,salinity to solonetzic soil. Ed JhonLibbeyeurotext, Canada p 8 -23.
SCHONFELD MA.,JOHNSON RC,CARVER BF and MORNHI NW, 1988 - Water relation in winter wheat as drought resistance indicators. Crop. Sci, p 28:528 -531.
SEHABI M., 2000 - Etude génétique et agronomique de la tolérance à la sécheresse chez quelques variétés d'orge (Hordeumvulgare L). Magister thesis. I.N.A.Alger, 92p.
SEMIANI M., 1997 - Etude de l'effet du stress hydrique sur quelques processus physiologiques et de croissance de deux variétés de blé tendre (Triticumestivum.L).Recherche Agronomiques. INRA A. vol (1) p 23 -32.
SHANNON MC, 1985 - Principles and strategies in breeding for higher salt tolerance. Plant Soi1; 89: p227 -241.
SOLTANI A., HAJJI M and GRIGNON C., 1990 - Recherche de factors limiting the mineral nutrition of barley in a saline environment. Agronomie, 10: p 857 -866.
TAL M., 1984 - Selection for stress tolerance. In hand book of plan cell culture, techniques for propagation and breeding (D.A evans,W.R.sharp,P.V.Amirato and Y. Yamadou, eds).Vol.1 p 461 -488. Mac Milainpublishingco.,New York- ISBNO.
TARDIEU F., N KATERJI and O BETHNOD, 1990 - Relation between water water status of soil the potential of low and other

indicators of water stress in maizeAgr:10 :p617 -626.
TORRECILLAS..1993 -Water relation and osmoticumsajustment in lycopersicumesculatum and lycopersiconpeunelle during short term salt exposure and recovery. Plant physio.(89) p.441 -447.
TURNER N C., 1986 - Adaptation to water deficits: a changing perspective - Aust-J.plant physiol. 13:175.
UNGAR IA., 1996 - Effect of salinity on seed germination, growth and ion accumulation of Atriplexpatula (Chenopodiaceae). Amer. J. Bot; 83: p 604 -607.
VAN-HOORN JW., 1995 - Development of soil salinity in the root zone N°2 séance spécilisée du 22 mars.p 65 -66.
VIEIRA SILVA, CLAUDIVAN F., HENRIQUE A., ENEAS J., ENEAS G., JOSE TARQUINIO P., 2003 - Physiological responses of NaCl stressed cowpea plants grown in nutrient solution supplemented with CaCl2. Braz. J. Plant Physiol. vol.15 no.2.
VILAIN M., 1987 - La production végétale, vol 01.Ed: la voisier, Paris agriculture d'aujourd'hui science, technique application, 416 P.
WHITTINGTON J and SMITH FA., 1992 - Salinity induced malate accumulation, Chara.J.Exp.Bot.43, p837 -842.
WILLIAMJ.,KATEMBE,IRWINA,UNGAR, JOHNP.MITCHELL, 1998 - Effect of salinity on germination and Seedling growth of two Atriplex species. Annals of Botany 2:p167 - 175.
ZAIR M., 1991 - Contribution à l'étude de l'influence du déficit hydrique sur le développement du blé en zone semi a ride revus céréaliculture N°24. Eds. ITGC. PP 9 -13.

APPENDIX

Table 13- Relationship between salinity and physiological parameters (Na Cl+CaCl2).

Salinity RWC1 RWC4 RWL TC PO1 PO4RsS1 S4
Salinity 1
RWC1 0.244 1
RWC4 0.003 0.5291
*RWL-0.707*0.0360.2471*

TC	-0.731*	0.192	0.086	-0.895**		1						
PO1	-0.588	-0.544	0.035	0.048	-0.179			1				
PO4	-0.918**	-0.191	0.189	0.587	-0.569		0.726*			1		
Rs	0.327	0.473	-0.204	-0.318	0.253		-0.254		-0.254			1
S4	0.690*	0.443	0.185	-0.549	0.674*		-0.566		-0.595		-0.006	1

Table 14-Relationship between salinity and physiological parameters (seawater).

	salinity	RWC 1		RWC 4		RWL TC		Rs S1	S4
salinity	1								
RWC1	-0.761*		1						
RWC4	-0.396	0.518			1				
RWL	-0.801**	0.545		0.184		1			
TC	-0.802**	-0.655		-0.239		-0.719*	1		
Rs	0.289	0.79		-0.137		0.123 .029 9		1	
S1	0.495	-0.382		-0.336		-0.568 0.677*		0.113 1	
S4	0.843**	-0.411		-0.386		-0.723* 0.689*		0.444 0.305	1

Table 15-Relationship between salinity, water status and morphological parameters of the root system (Na Cl+CaCl2).

	salinity	RWC1	RWC4	Long rac	PSA	PSR	PSR/PSA
salinity	1						
RWC1	0.244	1					
RWC4	0.003	0.247	1				
Long rac	0.259	-0.153	-0.351	1			
PSA	-0.762*	-0.014	0.198	0.208	1		
PSR	-0.752	-0.082	0.088	0.290	0.958**	1	
PSR/PSA	-0.241	-0.245	-0.245	0.611	0.451	0.669	1

Table 16-Relationship between salinity, water status and morphological parameters of the root system (seawater).

	salinity	RWC1	RWC4	Long rac	PSA	PSR	PSR/PSA
salinity	1						
RWC1	-0.761*	1					
RWC4	-0.396	0.518	1				
Long rac	-0.698*	0.197	-0.135	1			
PSA	-0.686*	0.417	0.240	0.720*	1		
PSR	-0.943**	0.698*	0.346	0.726*	0850**	1	
PSR/PSA	-0944**	0770*	0.399	.570	0.498	0.875**-	1

Table 17- Relationship between salinity, water status and leaf anatomical parameters (Na Cl+CaCl2).

	salinity	RWC1	RWC4	comp.meso	diam.xylm	long,celgain	larg,celgain	epméso,fvt	epméso,fds
salinity	1								
RWC1	0.244	1							
RWC4	0.003	0.247	1						
comp.meso	-0.442	-0.224	0.606	1					
diam.xylm	-0.213	-0.466	-0.355	0.355	1				
long,celgain	-0.329	0.107	-0.077	-0.102	-0.317	1			
larg,celgain	-0.493	0.008	-0.118	-0.026	0.094	0.731	1		
epméso,fvt	-0.474	-0.305	0.232	0.414	-0.213	-0.101	-0.260	1	
epméso,fds	-0.149	-0.381	-0.353	0.034	0.478	-0.117	0.058	0.107	1

Table 18- Relationship between salinity, water status and leaf anatomical parameters (seawater).

	salinity	RWC1	RWC4	comp.meso	diam.xylm	long,celgain	larg,celgain	epméso,fvt	epméso,fds
salinity	1								
RWC1	-0761*	1							
RWC4	-0.396	0.518	1						
comp.meso	0.111	-0.429	-0.628	1					
diam.xylm	0.296	-0.326	-0.695*	0.264	1				
long,celgain	0.405	-0.259	-0.355	0.419	0.85	1			
larg,celgain	0.720*	-0.298	-0.408	0.098	0.309	0.566	1		
epméso,fvt	-0.365	-0.024	-0.227	0.388	0.280	0.117	-0.030	1	
epméso,fds	-0.572	0.242	-0.322	0.096	0.490	-0.402	-0.383	0.520	1

Table 19- Relationship between salinity, water status and root anatomical parameters (Na Cl+CaCl2).

	salinity	RWC1	EPC	LCP	LACP	Dia v mx	No. of hpv	E rhiz
salinity	1							
RWC1	0.244	1						
EPC	0.019	-0.083	1					
LCP	-0.310	0.041	0.253	1				
LACP	-0.171	0.196	0.026	0.932**	1			
Dia v mx	-0.018	-0.092	-0.056	0.272	0.277	1		
No. of hpv	-0.401	-0.237	0.227	0.068	-0.212	0.315	1	
E rhiz	0.441	-0.012	0.296	0.295	0.363	-0.269	-0.499	1

Table 20-Relationship between salinity, water status and root anatomical parameters (seawater).

	salinity	RWC1	EPC	LCP	LACP	Dia v mx	No. of hpv	E rhiz
salinity	1							
RWC1	-0761*	1						
EPC	-0.500	0.544	1					
LCP	-0.543	0.699*	0.825**	1				
LACP	-0.297	0.474	0.799**	0.920**	1			
Dia v mx	-0.338	0.379	0.816**	0.631	0.663	1		
No. of hpv	-0.701	0.724*	0.822**	0.731	0.557	0.541	1	
E rhiz	-0.409	0.522	0.746*	0.671	0.689*	0.474	0.691*	1

MIX
Papier aus verantwortungsvollen Quellen
Paper from responsible sources
FSC® C105338

Printed by Books on Demand GmbH, Norderstedt / Germany